Footsteps in the Furrow

Footsteps In The Furrow

ANDREW ARBUCKLE

Sandy,

I hope you enjoy
these memories,

Andrew Arbuckle
July 2009

Old Pond Publishing

First published 2009

ISBN: 978-1-906853-05-1

A catalogue record for this book is available from the
British Library

Published by
Old Pond Publishing Ltd
Dencora Business Centre
36 Whitehouse Road
Ipswich
IP1 5LT
United Kingdom

www.oldpond.com

Cover design by Liz Whatling
Typesetting by Galleon Typesetting, Ipswich
Printed and bound in Malta by Gutenberg Press

Contents

Dedication

For Lydia and Elizabeth

Chapter 1

Introduction

As a boy, my walk to primary school took me along a quiet rural road running parallel with the south side of the river Tay from the small town of Newburgh. Along with half a dozen other youngsters, we would some days dawdle and play along the way. Other days, when the rain beat down upon us, we would scurry home as fast as our short legs would carry us.

Some sixty-odd years later, I still live along that same stretch of road looking out over the land my family farmed for the best part of half a century. The road is known locally as the Barony, after the Barons of Rothes, who, for five preceding centuries, owned the riverside strip of land. Their castle, Ballinbreich, now lies in a ruinous condition but still commands a dominant position looking upstream towards Perth and eastwards to the estuary of Scotland's largest river.

In those feudal times, the castle was the hub of all life. Small, unfenced bits of land might have been tilled around its sturdy walls. Sheep, and a few of the now-extinct Fife breed of cattle, would have been tended on the slopes. The purpose of the castle had little to do with defence against some invader. It was re-built in the sixteenth century but Fife had never been marauding country, leaving that activity to the more quarrelsome peoples in the Highlands and the Borders.

The castle had more to do with status. Its sturdy presence stamped its mark upon the area and also on the people who lived under the shadow of its walls in those days. Most of those living in the parish would be sheltered within the castle and tenant farmers paid their *feus* to the barons as they eked out a living from the land.

Although running roughly parallel with the riverside, the road takes the easy route, like all tracks born in the days of horse and cart. Hills were tackled gently, with no steep gradients; winding round the contours rather than heading for the shorter, steeper, more direct route.

It is a road where a steady pull on the cart shafts would transport the loads of grain and potatoes towards the local markets; a road where ridden horses could also keep steady pace without breaking stride to cope with sudden ups and downs on the carriageway. To call it a 'carriageway' is somewhat grand. It was a statute labour road, meaning the adjoining landowners were required to carry out the maintenance on it.

This was never a main road between two important points. In the early days, the Barony road would have been no more than a couple of stone-filled tracks for the cart wheels to follow and a softer, unmade up section between for the horse. Only in the early days of the twentieth century did the local authority get round to covering it with tarmacadam, classifying it in their bureaucratic way as 'C46'.

In those days, the main town of Newburgh had a corn market to which grain merchants from Perth and Dundee would travel, either by horse or by boat. Grain and potatoes for markets in the south of England were loaded onto boats by the simple expedient of horse and cart backing down towards the vessel that lay beached at low tide. Later, in my time, farm produce was transported by tractors and trailers, hauling seed potatoes from the farm towards the station in Newburgh and then onward to their destination in the south of England.

Horse carts and gigs have long gone and although the main Edinburgh to Perth line still crosses the land, the railway closed down four decades ago in Newburgh. Today, agricultural traffic consists of large articulated lorries and is largely limited to a two-month period at harvest time. It sees bulk lorries of grain with 20-plus tonnes of wheat or barley, heading for the malting or distilling markets – or, if the quality is less than it should be, for feed mills. For an equally short period, unwary rural travellers may also encounter large heavy goods vehicles with potatoes in 1-tonne wooden crates being driven away to centralised stores.

The road still winds through the countryside, but in the past hundred years farming has changed more dramatically than in a score of centuries. At the beginning of the twentieth century, along the 5-mile length of the Barony road were a dozen farms. Some were small, with only the tenant working the acres and keeping a few cattle and sheep. Ownership of the estate passed to the Zetland family – absentee landlords, who acquired it from the Rothes family through a marital link.

Some fifty years later, in the middle of the twentieth century, the number of working farms had shrunk to 7. Smaller units were subsumed into the larger ones, to leave only a small biggin – a small set of farm buildings – behind.

On these 7 farms were some 25 cottages, with another 4 added by the estate in the 1950s to cope with demand for additional farm workers. The valuation roll of those days showed all 29 dwellings housed farm workers.

As we march onward in the twenty-first century, only three of the original dozen farms work as independent units. The rest are farmed from outside the parish and all the arable work is done in a short burst of feverish activity at springtime and a slightly longer bout of high-tempo activity in the autumn. One solitary farm along the road has retained its livestock enterprises, so it is still possible to see newborn calves from the commercial suckler herd.

In the spring, the latest crop of lambs can be seen initially looking as if they want to confirm all the prejudices farmers have about sheep having a death wish. However, within a few days, they, and dozens of their colleagues, romp about the fields engaged in pointless, but joyful chases. The rest of the livestock along this parish road arrives for the summer grazing season and departs either to market, or back to the owner's farm, several miles away.

Some twenty-eight of the cottages remain, but not one of these provides accommodation for a farm worker. With one exception, work on the farms is carried out by the farmers themselves, or by contractors coming in with seeders or harvesters. The one remaining full-time agricultural worker lives in Cupar – some twenty years after he left the tied house in which I now live.

All the cottages and the parish school and church are now occupied by those who commute from their rural base into the towns and cities. In contrast with the tied housing of a previous generation, many are now owned by those who live in them.

The local primary school has been closed for the past thirty years, and today's children are collected and deposited, morning and night, by a taxi service that takes them six miles to the nearest rural school. The church has also closed its doors. It sits, roofless, surrounded by a graveyard. Headstones tell of the farmers and workers of previous years. For most of the day, this quiet country road is almost deserted, following the early-morning dash to work until the return journeys in the evening.

Recording the changes

Some twenty years ago, when I gave up active farming to write about agriculture in the Dundee *Courier*, I would occasionally refer back to farming practices of previous generations. Descriptions of the machine harvesting of potatoes compared with the hand picking of the crop would bring forth letters full of such memories. Similarly, comments on how the Scottish summer raspberry crop would largely be picked by holidaying Glaswegians resulted in phone calls, recalling those seemingly halcyon days. Even reports on relatively mundane heavy physical work, such as dung spreading, seemed to provoke fond memories.

Although the husbandry learned more than half a century ago is still relevant, many of the skills gathered at that time, when labour was an essential of good farming, are no longer part and parcel of farming life. For example, being able to mark out 'bits', as the sections of the field were called at potato picking time, lies in the basket of skills now laid to one side and labelled redundant. Likewise, an ability to measure out the capacity of a straw stack is an attribute that now moulders away in the recesses of a few elderly minds.

Apart from the skills skittering away from modern man's brain cells, many of the customs and practices linked to farming in the last century have disappeared. Gone are the days of labour hierarchy on the farms. With no farm grieves (working farm managers) and no orramen ('ordinary' farm workers), this ladder of rural social life has lost its rungs. Gone is the chat or crack between the team of men on the farm and the loon, often a callow youth who was always on the butt end of any prank, or joke – such as sending him for a load of postholes or a tin of tartan paint.

This book is an attempt to shine a light on life on farms in the previous century and to capture some of those work practices and pictures of a rural landscape from yesteryear. It does not pretend to be a history of farming. Although the major events shaping the industry are recorded, they are there merely as directional markers, not part of a definitive history. Nor does it have any pretensions to be a sociological record of rural life: that would be too grandiose an ambition for what is no more than a collection of memories.

The great temptation when looking back into the past is to forget the downside to a simpler way of life and to remember only the good parts. While everyone remembers the camaraderie, few will talk of the harshness of life, where a wage earner's illness or accident would quickly leave a family clinging onto the proverbial bread line. And while there are happy recollections of harvest fields full of workers, the reality of those days was also one where working conditions were often unpleasant, sometimes severely so.

Pictures of rows of workers standing outside the stables holding their horses reflect the pride and fellowship of the work in the era of the dominant horse. But the photographer was not on hand when those same men were out ploughing in the sleet and rain; sometimes sheltering under the horse with just with an old sack over the shoulders to keep the worst of the weather at bay. I hope that in these memories a fair approach has been taken, one that recognises that some parts were good, especially the camaraderie, and others were not so great.

The physical boundaries of the stories are mainly kept to those around my calf country, that of North-East Fife, but there is a well known saying in the Scottish farming industry and that is, 'If you want to see the whole country but do not have the time, then just go to Fife.' It may be one of the most clichéd descriptions of the county, but calling it 'a beggar's mantle fringed with gold' describes the rich, fertile coastal strips surrounding the slightly poorer land in the centre.

I have also no doubt that many of the practices recalled within these pages have similarity with those from other areas. The geography is no more than a sampler onto which memories and stories are stitched. One further qualification: this is not a personal history, or even a history of my own family. In farming terms, the Arbuckle family was no different from many others in their origins and work. To their cost, they might have dabbled more deeply in the politics of farming than most, but that is not part of the story.

My own little store of memories and family records has been greatly augmented by the many kind friends who spoke openly of their recollections of times gone past. The verbal harvesting of customs and practices of the older generation has been one of the joys and happiness of this work.

I hope your reading of this book will either tug at your own personal memories, or, if you are of a younger generation, provide an insight to how life used to be down on the farm.

ANDREW ARBUCKLE
Newburgh, Fife, 2009

Chapter 2

Early Days

MY grandfather, John Arbuckle, was brought up on the family farm on the outskirts of Bathgate in the industrial heartlands of Scotland. He was reputed to have married one day, and the very next morning to have taken his new bride and all the possessions essential to taking the tenancy of a farm off in a horse-drawn cart to their new home in Angus.

There is no record of the length of time this journey took in the first decade of the last century but it would have retraced the steps of the drovers who, in the 100 years prior to that, brought cattle and sheep down from the hills and glens to the big Tryst at Falkirk.

The new couple set up home on a small farm outside Glamis in Angus, where they lived and worked for a number of years before a bigger tenancy came along.

And that was how my family came into the county of Fife. The slightly circuitous route may have been unusual, but all through the first half of last century, there was a tidal flow of new farming blood coming into Fife from smaller family farms in the wetter west of Scotland. Such was the scale of this migration that by the 1950s there were very few farmers in Fife unable to trace their roots back west. There were a number of reasons for this, but the biggest single factor lay in the economic benefits from the increased options provided by farming in the east of Scotland.

The east-coast climate allows a wider range of crops to be grown and the soils are generally better than the wet, heavier land in the west. Grass has always grown well in the west, but as Scotland's national poet, Robert Burns, knew to his cost, growing crops in Ayrshire was not an easy option.

Those migrating from the west did not all come east; many went south, where the land was again of a better quality and the climate drier and warmer than their homelands. The nature of migration is such that the successful pioneers who blazed a trail into new territory tended to get the message back home. This sets up a second wave of migration and more hopefuls head towards the Promised Land. To this day, there are farming villages in Lincolnshire, Essex and Bedford where those of Scottish origin dominate.

Although not used by my grandfather, the migration into the east was made much easier by the network of railways that criss-crossed their way across the country. It was possible, with a little planning, to move entire farms – livestock, goods and chattels – by train. Many of those coming eastwards would milk their dairy cattle in the morning in their old farm in the west and then carry out the afternoon's milking at the new base. Some hired whole trains to carry out their flitting. When John Steven came to Stravithie, outside St Andrews, his bill from the railway company for moving the lock, stock and barrel of his farm across Scotland was £39 10/-. This might be multiplied by twenty times to cater for a century of financial inflation and still it would be a tremendous bargain.

Another family making the same move eastwards were the Logans, who came to Dairsie Mains complete with their herd of Ayrshire milking cows. Again, those cows were hand-milked in the west before going on the train, and as they settled down in their new home, they were milked in the afternoon. The Logan family also demonstrated considerable acumen as they sold turf from their new farm to help create the world-famous golf course at Carnoustie, and with the proceeds they bought another farm: Kirkmay at Crail.

Even with the advantage of coming east to farm, anyone regarding it as the Promised Land must have been labelled optimistic in the early years of the century. For the previous three decades from the 1870s, agriculture had slipped into a deep depression. Long gone were the golden days of the earlier 1800s, when the protection afforded by the Corn Laws – import tariffs designed to support domestic British corn prices against competition from less expensive foreign imports – produced prosperity never before or since seen in the countryside.

During those years of plenty, investment in agriculture transformed the landscape. Many of the farm steadings in the country were built around this time to house livestock, store crops and shelter machinery. Drainage and the general improvement of soils through liming and marling (adding clay to improve light, sandy soil) also came about in this period of abundance.

With the repeal of the Corn Laws in 1846 and the subsequent surge of imports to help feed the newly industrialised population in the UK, the economics of agriculture within those shores collapsed. As the rich and massive prairie hinterland of America opened up and the plows – American spelling – cut open the vast grasslands, once roamed by buffalo, imports of grain began in large quantities, thus destroying the home market in the 1870s. As a direct consequence, the acreage on corn in the UK fell by 25% in the 20-year period between 1873 and 1893.

From the other side of the globe, imports of mutton

flooded in from New Zealand and Australia after it was found to be economically possible to ship meat for as little as two pennies per pound, or 2p per kilo. Even when the cost of transport and the cheaper initial price of the lamb were added up, the total was still far less than the home-produced product. A similar story of cheap beef from South America undercut the UK market to the extent that numbers of beef cattle in the UK were 20% fewer in 1900 than only ten years previously.

In the final decade of the nineteenth century, the consequence of all those cheap imports was that prices were halved. One quarter of the agricultural workers left the land in that same period. Farms fell into dereliction or were reduced to grass as the practice of growing more expensive crops was abandoned. Many tenant farmers simply vanished, leaving the landlord with empty farms.

As they entered the twentieth century, farmers in the UK recognised that their fortunes depended not on how they managed their own costs, but on how their government controlled imports. This position has remained constant throughout the last 100 years and will persist in the future.

As will be seen, only in times of shortage or privation do governments place importance on the home production of food. At all other times, the benefits of cheap food to the wider economy dominate the political thought process.

During the pre-war hard times when they were getting low prices for their produce, farmers did what farmers could do better than most and that was to tighten their belts and cut costs. This rule applies to the present day and comes into play whenever the economic barometer hits the floor. Often it sees land, once ploughed, revert to being grassed over, with livestock running extensively. 'Dog-and-stick farming' it is called, employed by many a successful farmer for his survival. Reducing the amount of money going down the farm road and off the farm may not have been the recipe for high-living, but often ensured the ability to continue living on the land.

It was reckoned that for many the main outgoings on many of the farms in the first decade of the century were limited to paying blacksmith's bills and the small pittance that workers received for wages and a few essentials, such as clothes and food. Even the food that was bought was often obtained under a barter system, with eggs and butter exchanged for salt and spices that could not be home produced.

Even with self-imposed austerity, some farmers could not see their way through the Depression. One such example saw the disappearance in 1912 of Thomas Hunter White, the tenant of Drumrack farm outside St Andrews, halfway through his agreed tack, or term of lease. A local banker, Henry Watson, received a letter from an advisor in the Edinburgh and East of Scotland College of Agriculture telling him of the situation. Mr Watson, who had always wanted to farm, asked for a description of the farm. Back came the letter:

> *The whole farm is deplorably dirty. It is ostensibly down to grass but it will require to be sown again. The steading is antiquated and has fallen into a state of disrepair. The cottages are bad and would require to be gutted. The farm has been culpably neglected.*

Mr Watson took the farm and now, almost 100 years later, his family still farms the same unit. Possibly unwittingly, he had chosen a period when farming was moving into a short period of profitability.

What many farmers and politicians did not know or did not recognise in the first decade of the last century was that

Britain would shortly be engulfed in the largest conflict ever experienced: World War I.

To prove that modern politicians do not have a monopoly on daft statements, in 1903 the Board of Agriculture, the then equivalent of the Ministerial Department responsible for the farming industry, produced a report on food supplies. It stated: 'There is no risk of cessation of supplies, no reasonable probability of serious interference with them and even with a maritime war, there will be no material diminution of their volume.' Little more than a decade after that imperious statement, German warships sank shiploads of wheat, oats, beef and lamb destined for Britain.

At the start of the war in 1914, more than half the food consumed in the UK came in from abroad. Five years later, at the end of the war, British farmers were producing two-thirds of the food for a hungry home population. And they did so profitably. Even if rationing helped prevent the worst excesses of profiteering, farmers benefited from price increases and ready markets for their produce. Andy McLaren, of Nether Strathkinness, St Andrews, will be among the last of a generation to remember the latter years of that war being hungry ones, with food rationing having to be brought into play.

That war, with prowling enemy ships sinking boats full of food coming into Britain, first sharpened the nation's attention to food security. Whatever words are used, it was the war that brought home to this island nation the fact that importing food during conflict brought hazards, such as shipping supply lines being cut by the enemy.

Even before the last guns of war sounded, the then government was planning legislation that would provide a guaranteed price for wheat and oats. It was 1920 before the Agricultural Act actually came into being, with its promise of a minimum price of 95 shillings per quarter for both wheat and oats. But within months of this financial protective shield, food imports again swamped the country. Much to the annoyance of the recently created National Farmers Union of Scotland, the government, faced with high post-war unemployment, decided to ditch the Act as a quick way of lowering the price of food.

The vast majority of farming in Fife in the early years of the last century was carried out through the tenancy system. Even though there were no large-scale landlords in Fife, such as can be seen throughout the rest of Scotland, many private landlords in the county had small, tenanted estates. Fife landlords had been characterised by the saying that they 'had a wee puckle land, a doo'cot, and a law suit', which translates into their ownership of a small acreage of land, a pigeon loft for food and a penchant for argument. In 1908, some 90% of the farmed land was tenanted. That percentage was stable until the end of World War I, when cash-rich tenants took advantage of cash-strapped landlords and bought their farms.

So, why were the landlords short of cash? Simply because leases were taken on a 7- or 14-year basis and landlords were unable to raise rents during the good times; reportedly, rents were actually less at the end of the war than at the beginning.

Over the next fifteen years, the percentage of owner-occupiers rose to more than 30%. It was the biggest shift in land ownership in Scotland since monastic times. The change of land ownership was helped along the way by the increase in Estate Duty, introduced in 1925. However, many of those tenants who had taken the leap into ownership may have regretted taking that step as farm commodity prices plunged in the early 1920s to levels last experienced ten and twenty years previously. Other factors further accentuating the post-war economic downturn came into play. The savage loss of life in the war had robbed the country and the farms of a large percentage of its workforce. Some indication of the Grim Reaper's scything down of young men belonging to the rural areas can be seen to the present day just by reading the war memorials scattered around the countryside. As more and more young men were called away to fight for King and country, it was not unusual to see boys as young as 14 years of age ploughing behind a pair of horse.

Fewer men were left on the farm and those who did remain required more pay, but when commodity prices plummeted, farmers tried to negotiate a drop in wages. As we will see elsewhere, these negotiations helped form the Farm Servants' Union.

Similar economic woes hit the urban areas and farming was in recession more quickly than was at first realised in the 1920s. In a perceptive remark in 1929, Mr Henderson of Scotscraig, the president of Cupar NFU, stated that because farmers were not getting prices to meet the costs of production, large tracts of land were going down into grass: 'Certainly, the Nation will awake to this fatal error but then it may be too late with the farming interest largely wiped out.'

In fact, the nadir for the farming industry came in the early 1930s. The average area of land under cultivation in Scotland during that decade was the smallest since 1876. Even in the run-up to World War II, production of all commodities except poultry was much less than at its peak in 1918. This happened despite the efforts of Walter Elliot, a Scottish sheep farmer who became the Minister of Agriculture in the early years of the decade. By imposing some import tariffs and encouraging farmers to grow for defined markets, he tried to bolster the industry.

In the 1930s, there were many cases of bankruptcy and

then there were the less publicised ones of suicide. It was said at the time that no farmer could look out over his neighbourhood without seeing at least one farm where the farmer had either taken his own life or had simply vanished from the scene. In the depressed years, the small parish of Carnbee lost two farmers who took their own lives, while another six went bankrupt. As was witnessed in an earlier era at Drumrack, farms let on tacks or rental periods of 7–14 years regularly were abandoned by the tenants.

Far more than in any other occupation, failure in farming, for whatever reason, is a heavy burden and one that is often carried alone. However, it was in those deep, dark days that the industry first pulled itself together and set up organisations to fight its corner. Their story is carried elsewhere in this book.

Chapter 3

Farms, Fields and Steadings

WHEN travelling through towns nowadays, every so often you come across a children's play area, filled with swings, chutes, roundabouts and climbing frames. They are in stark primary colours and mostly surrounded by a soft rubberised area, lest the children fall off. It all seems so distant from the play area of my own childhood and of those brought up on the farms of a generation or more ago.

For youngsters growing up in the country, farmyards were our playground – the buildings, the farm machinery and the goods stored about the place provided the landscape for rich adventure and risky escapades. One favourite starting point was the farm stables; the horses had gone from the farm but the stables in the farm steading still had a certain relevance.

The size or the number of stalls in a stable gave an idea of the scale and nature of the farm. A rule of thumb was that a pair of horse could cover some 50 acres on an arable farm and so a quick count of the number of stalls would tell the size of the farm.

Some farms in Fife had two stables. Unusual, until you consider that in World War I, often the military came onto farms, unannounced, and requisitioned the best of the horses. Farmers found it was far better to split the risk and hope that those responsible for taking away the horsepower of the country to the battlefields would not realise they were only seeing half the horses on the farm.

From an early age, we youngsters could open the half-hack stable doors, where the top half could be opened for ventilation, leaving the bottom still closed with a latch or a draw bolt. Once inside the stone-built building, we could climb up onto the food troughs at the front of the stalls, which were separated by wooden partitions. Even by this age our senses were heightened by the sharp smell of the creosote applied to every piece of wood in contact with livestock.

We did not know it then but this form of disinfection was applied on an annual basis to keep at bay contagious diseases such as ringworm. Nor did we realise at that time that there was a similar reason for whitewashing all the stone and brickwork with limestone. For some, white-washed walls equate to cleanliness and good hygiene.

For myself, and others who applied the hot lime or the creosote during the summer months when the livestock were outside, there are memories of stinging faces whenever the brushes accidentally splashed preservative onto our skin. This routine work was not enjoyed. In addition to the discomfort experienced by wayward splashes and drips, working clothes were marked. There being no overalls in those days, the general practice was to cut a hole in the base of an old jute sack and then stick your head through it, leaving the wearer with the latest fashion: a bag advertising Bloggs Grain.

From the food troughs, with a slightly acrobatic swing we could get into the slatted hay haiks above the troughs. On some farms, there was a direct connection with the loft next door so that hay or oat straw could be fed through the boles. Crawling through these boles would then take you into the loft with its shiny wooden floor; smoothed and ribbed by years and years of forage being dragged along as it was filled and gradually emptied.

On some farms these lofts would be filled by straw conveyers slung from the roof rafters straight from the threshing mill; others had conveyors that took the bulk grain into the dry feed store that was also filled with sacks of various foods for livestock. In both, bags or bulk, there were a thousand dusty hidey-holes where youngsters might conceal themselves when playing hide and seek.

The loft was accessed by a set of stone steps. These steps were worn away by the tackets, or short nails in the soles of the boots of a generation of men who carried sacks of grain weighing more than 100 kilos. Pre-Health and Safety Executive days, there were no handrails on the steps and so, if brave or foolish enough, we youngsters could jump off the top step onto the cobbled yard below.

Below the loft was the cart shed where the coup carts, and later small tractor trailers were housed; each with its own beautifully constructed archway of stonework built in an arc with a keystone. At the highpoint of the arch was a hook, on which the carter would hang the shafts of the cart. Alongside each cart were hooks on which the extra cart sides were stored, in case the work involved a bulky crop. Above, on the wooden rafters were the flakes used when carting hay and straw.

At the other end of the loft were the cattle courts and again, we, as children out of sight of adult restraint, would climb into the troughs running along each side of the raised gangway. We would then climb into the hay haiks and up onto the couples, or roof rafters. From this high vantage point, it was a test of courage to crawl along these wooden batons looking down on the cattle below. Then, in a game of dares we would hang from the beams until an unsuspecting bullock wandered below. The aim was to land on its back, but more often we missed and fell onto well-trodden dung.

As we stepped from rafter to rafter we had little thought of the joiners of a previous century possibly having skimped on their nailing. In our escapades, we believed that accidents were for other people; any superficial damage would sort itself. That is why short trousers were always worn because skinned knees were cheaper to sort out than tears in long trousers.

Below us were the cattle that were spending their winter being fed and watered as part of the fattening process. Fife was, and still is, an area for finishing cattle and winter housing was required for this purpose.

On farms with breeding herds there would be some smaller buildings with stalls, where the cattle would be tethered by the neck. These buildings had large, vertical flagstones separating the stalls. There was a food trough in front and a dung passage at the back, where the day's animal waste, as we never called it, could be swept along to the end of the shed before being barrowed out to the midden – the heap of waste and animal dung.

Milk cows were also tethered in the same type of stalls so that milking could be carried out in relative safety with only the danger of a kick from the hind legs, if the cow did not relish the milking process.

When we tired of the cattle courts or the milking stalls, with their warm, moist, sweet smell, we would play in the turnip shed, which was handily built next door so that the cattlemen had only short trips to make between shed and trough. The turnip shed was less fun and invariably led to excursions onto the roofs of the steading itself; the route was through a broken roof light, then a clamber up the pantiles, trying hard not to dislodge them.

Again, we seemed to care little for the robustness of the

roofs. We could see down into the sheds, but never thought that the whole roof might be somewhat unsafe. The brave walked along the ridges, from where they could see the whole layout of the steading, all the time hoping no adult would see them.

Many of the old farm steadings were built in a U-shape, with the farmhouse often helping to fill the gap in the 'U'. From a high point on the roof, our eyes followed around. First, the stables, then the loft and cart sheds, and onto the cattle courts, always bounded in by the turnip shed.

Many of the smaller farms had a horse mill. This was a separate hexagonal building, in which the power to drive the threshing mill was generated by a single horse pulling a shaft that drove a central hub or capstan. The old steadings were built for, and by, horsepower, though on the larger farms steam engines may have puffed away, turning the wheels of the threshing mills.

The majority of the steadings in the arable parts of Fife were built in the middle and late 1800s. Most of the farms were tenanted and landlords, keen on improvement, built farm steadings for their tenants. Stone was the main material used in the construction and the buildings were built of local sandstone or harder whinstone. Because of the cost and effort of transporting stone, many quarries were created purely to supply material for farm steadings and cottages.

In that busy building era, most of the parishes had several stone masons. The buildings they created reflect the agricultural priorities of the area, as well as the relevant importance of both the farmhouse and the farm cottages. But even in my youth, there were additions to these traditional steadings, thus proving the old adage that no farmer, however intelligent he claimed to be, ever built his steading big enough or his field gates wide enough.

The latter point related not just to the ever-increasing scale of farm machinery, but also to the fact that generations of ploughmen, farmers and farm students have notoriously been unable to guide a tractor or implement into, or out of, a field without touching, scraping, or even the downright breaking of a gatepost. The arrival of the tractor saw great ugly holes being punched through the original stone walls as the old stable door had not been built wide enough to allow access for mechanical vehicles.

Some of the earliest additions to farm steadings were former World War II buildings, which were given a second lease of life as implement sheds, henhouses or pig-fattening buildings. Many of these were made of corrugated iron; others were pre-fabricated buildings.

On some farms, silage towers had been built in the 1920s and 1930s. These were often made with concrete and a few examples such as the one at Collairnie Farm, Letham still exist. Then, as new materials came along and more knowledge of silage making came into use, fibreglass sealed silage towers soon pierced the skyline. With less demand for grass-based forage, there were always fewer of these in Fife than in dairying districts of Scotland.

The boy on the roof of the old buildings in the 1950s could also see the first bulk grain bins built close to the steadings. These came in with the combines when lifting heavy sacks fell out of favour. Conveying grain electrically by auger and elevator was found to be far more efficient and much quicker.

The first of the specialist potato sheds also came into being in the post-war years. These were brick-built, with asbestos sheeting over steel trusses. As technology advanced, later models dispensed with the trusses, replacing them with steel beams. This allowed farmers to maximise storage space by stacking the potato boxes higher than previously imagined.

Today's modern potato shed comes with ambient temperature control that removes the old problem of tuber diseases spreading through the crop when the potatoes overheated after being lifted in wet conditions. It also stops the sprouting of potatoes in warmer weather.

As husbandry knowledge increased, specialist livestock buildings were erected. This was especially true for pigs and poultry. Long, low buildings with controlled ventilation first went up in the early 1960s. Outside these were metal feed hoppers to automatically feed the livestock; inside, the intensive production of poultry or pig meat was carried out.

For those farms still considered working units, the footprint of the buildings has multiplied several times during the course of the last century as crop storage and livestock production moved indoors.

At the start of World War II, there were some 1,243 farms or landholdings in North-East Fife. Today's total of viable working farms in the same area numbers less than 300. What has also happened is a congregation into fewer, but larger working units. Quietly, many smaller farms have been taken over.

Often no agricultural use is made of the farm buildings on these smaller units. Many, especially those around St Andrews, where there is a strong demand for accommodation from those working or studying in the ancient university, have been converted into housing. Once a month, the Planning Committee of North-East Fife area of Fife Council meets in Cupar. Almost without exception over the past decade, in the normal list of planning applications there have been bids to convert redundant farm steadings into housing on a regular basis. These are invariably granted. At least in this local authority area there is a requirement that any conversion largely takes place within the original curtilage or footprint of the buildings, with as much of the original building as possible retained.

Other councils take a more relaxed view. They allow old farm buildings to be demolished and then transplant a clutch of largely identical houses or a small piece of suburbia onto the flattened site. Even containing any development into the area previously covered by cart sheds, barns, lofts, cattle courts and neep sheds, sufficient space can be created for a dozen or so houses. So, now, theoretically, we have a repopulation of the countryside.

The reality is different as there is little or no connection between the work of the land and those who live in the steading conversions; the vast majority drive to work early in the morning and return late at night, the week's shopping achieved at some distant retail park.

It is fanciful to think that the ghosts from the past inhabit these old buildings recycled into modern homes; it is difficult to believe today's inhabitants, looking out from their floor-to-ceiling windows placed in openings of the old cart sheds, hear the voice of the old grieve shouting across the close about some perceived failing by one of the loons. And as they rush out to their cars to go to work, they will never hear the clip-clop of horseshoes over the cobbles at the start of day, or the sound of the turnip, or neep hasher, getting the daily diet for the cattle.

The past century has also seen a loss of farms and buildings directly as a result of towns and villages expanding their boundaries. Those picking up their ancestral roots often come back to the family farm to find they require to walk the concrete pavements or muddy playing fields of the burgeoning urban landscape.

The scale of change in farming is not easily observed

from the traditional view over the hedge or dyke, or even through fence wires on visits along the rural roads. The fields provide permanence, and crops are still grown.

Often, if the rural jungle drums do not beat out the message, the first that even neighbours now know of a change in working the land may come with the arrival of a stranger's set of machinery entering the fields. Gone too are the smallholdings, including those set up specifically after World War I by the British Government in their repatriation of soldiers – part of their belief that they were creating a land fit for heroes.

In North-East Fife two larger properties were split up to make smaller holdings – Third Part and Easter Pitcorthie. These holdings each had a farmhouse and steading, and approximately 50 acres (or 20 hectares). Today, only two of these original holdings remain, the rest have either been amalgamated or the land sold off to neighbours, leaving a house in the country.

Fields
The old trick of looking at the placement of a gateway to see who owned the field has also gone with the aggregation into larger units. It used to be that the gate was always placed in the corner nearest the farm steading. That was the shortest route for the horse to walk and for any work to be done. Nowadays, with a takeover of husbandry, that trick no longer tells tales.

The youngster who in the mid-1950s perched precariously on the old steading roof, would have been able to look beyond the immediate buildings to see the stackyard, possibly even a pond for water power and then a scutter of henhouses in nearby fields, or even a small paddock in which the tups (rams), or some ailing animal would be kept.

Beyond the buildings and the in-bye enterprises were the cropped and grazed fields. In the early days of the century, when the average size of a farm in Fife was 112 acres, seldom would the field size go beyond 20 acres.

There were still areas of land unfenced up in the rigging or highlands of Fife, but the vast majority of farmland was enclosed. A century previously, it was reckoned that only one-third of the land in Fife had been fenced or hedged into small workable fields. But even if they were enclosed, some fields were in a fairly basic state. In 1917, Mr Watson of Drumrack Farm, Anstruther paid the rector of local Waid Academy some £7 7/6 (£7.37) for the use of a squad of boys to clear the field of whins (gorse). At the same time, he also employed a team of Waid Academy girls to clear up the fields. This latter task was most likely one of hand-weeding crops such as potatoes or turnips, but it might also have included taking weeds out of grain crops.

Following massive investment in this unseen aspect of good husbandry during the previous century, the majority of the improved fields were also drained. Originally, the drains were trenches into which stones were placed, thus allowing water to flow between them. That rough description does no credit to the quality of the stone drainage work that still operates after more than 100 years.

Anyone who has had to repair a stone drain will confirm it is much more difficult than replacing a tile drain. These clay tile drains came into being on land where there were few stones. Generally it was an easier system laying these hollow cylindrical tiles, which – provided there was a run or gradient – would work effectively. Most of the drainage work was carried out in herringbone systems, with leader drains forming the spine and these emptying into open ditches or streams.

Fife is not an area where hedges are common, although 200 years ago the most popular method of creating fields was the combination of ditch and blackthorn hedge. There were also areas where stone was plentiful for building dry-stone dykes, but again, this was not a widespread practice as there are parts of Fife that are stone-free and, in horsepower days, the carting of stone was costly.

Not until fence wire became popular did the enclosure of farms become complete. As an example of the cost of fencing in the early days of the century, at Drumrack Farm outside Anstruther in 1914, Mr Watson paid £38 for the post and wire fencing of 1,000 yards, or just over 900 metres.

One of the first industrial imports from the US was barbed wire, used extensively and controversially to enclose the vast prairie ground. But it did not arrive without its problems, as in 1921 the NFU of Scotland received a communication from the Ministry of Agriculture and Fisheries reminding farmers of the injuries that could be sustained by indiscriminate use of barbed wire. However, barbed wire is still, to this day, an essential part of any new livestock fencing.

Most farm gates were made of wood and, equally, most were just tied or roughly hung on the wooden gateposts. Only the major estates or home farms could afford properly hung metal gates. Almost all these early gateways have been demolished as the width and scale of the farm machinery has increased and now there is often just a large, open gash in the hedge, fence or dyke to allow the machinery access to the field.

In a slow, but constant process since World War II, there has also been an increase in the size of fields as fences, hedges and dykes have been removed. In some arable parts of the East Neuk, there is no fencing at all, but as one farmer remarked, 'You do not need a fence to keep the potatoes in.'

Another slippage from the scene is the loss of field names. Old maps may still hang on farm office walls, but the names of individual fields have gone. These ranged from the self-evident 'Quarry' field, where a big hole would signify where the stone used in local building and possibly even for the farm steading had been quarried.

Fields with names such as 'Stoney Knowes' would no doubt give the ploughman thought as to exactly where the rocky outcrops might make his life difficult. Some field names gave the game away as to their previous history. 'Coal' field at Brigton Farm, St Andrews, was once worked for coal. Most people know of the coal-mining industry in west and central Fife, but right up until the late 1940s there were coal workings in east Fife.

The field on another farm called 'Clay Pit' may well have been the source of the pantiles on the roof of old steadings, but equally offered little in arable cropping. A field known as 'Holly Hedge' would again be named for obvious reasons and further down in the lower ground were the fields named 'Burnside'.

At Logie Farm, Newburgh, there was a field called 'The field with the stone in it' as it had a massive boulder, around which all farm machinery had to dodge. A visit from the local quarrymaster, along with a heap of dynamite, changed its name to 'The field without the stone in it'. Every farm had a 'Big' field just as every one had a 'Paddock'. A few even had a 'Big Paddock' and a 'Small Paddock'.

Few farms went to the same lengths as one owner at Falfield Farm, Peat Inn, when he erected a pillar made of sandstone specially brought down from Arbroath at each field. Each pillar proudly carried the field name.

So, what was the point of all the field naming? Well, it was easy for the farm grieve to tell the ploughman to go to plough the 'Big' field, or to manure the 'Big' paddock. Errors were known and men were found in the wrong fields, starting to plough where no cultivation was intended that year, or applying fertiliser when the required amount had already been broadcast. That was why field names were important.

Field names also gave an identity only taken away by scale and bureaucracy. All today's civil servant checking the forms filled in by the farmer needs to know is the Ordnance Survey field no. 123, on farm code no. 456, and he or she can then check by satellite the actions on that field.

Before leaving the naming and breakdown of the landscape into small parcels, it is important to mention that every parish would have had its church, and every church its own land, or Glebe. Some were quite sizeable, with the Glebe at Cameron amounting to 24 acres. While in the early years of the century many ministers would use this land for the grazing of their horse, this tradition slipped away by the middle of the century. Nowadays, most of the Glebe land is let to the nearest neighbour.

Chapter 4

Workforce

DRIVING home the message that my brothers and I were fully aware that money did not grow on trees, my father always ensured there was work to be done before he handed over any cash. And that was why, one summer holiday, I was set the task of painting the 'tin shed' on the farm. This was a straightforward structure, open on one side, to allow access for the machinery to be stored inside it. It was constructed of corrugated iron sheets that gave it a semi-circular roof and this was the object of my paintwork. The only trouble was that by the end of the day there seemed to be as much paint on me as the shed.

Immediately next to the tin shed was the bothy, a square wooden hut that was home to two Irishmen who came to work at the harvest and the sugar beet. One of them, seeing the state I was in, offered to help clean me up a little before I went back home. Shortly afterwards, I was sitting in the bothy with Paddy – whether or not this was his real name, I will never know – as he wiped the paint off my face, using one of his old socks dipped in an old jam jar filled with petrol.

Seeking to distract myself from the burning sensation on my face, I looked around the single-room building. There was a fire to one side of which a kettle was boiling, and an iron grid on the other side to be swung over the fire with a cooking pot. The beds were two single bunks, one above

the other, with grey blankets hanging over the side. I did
not see the mattresses, but they would be filled with chaff –
the common source of bedding on the farm. Obviously, I
did not see what we schoolboys called 'loupers' or bed
bugs, although in some bothies these little biting beasts
became a real scourge. Close to the bunks were several
clothes hooks, on which the Sunday clothes hung.

My body-paint remover and I were seated at a table
covered in newspaper. On the table were a loaf of bread, an
open jar of jam and a tin of meat paste, and that appeared to
be the only food available. It was pretty basic living, even for
the early 1950s. Water was collected from the tap that fed
the horse troughs. And the toilet? Well, I never thought
about it then, but it must have been in the cattle courts.

Farm bothies have been part of the folklore of Scottish
agriculture and in some parts, such as Aberdeenshire, a
culture was built up around them and the men who lived
in them. However, the bothy system was not always seen
as a good thing, and in 1891 a government report into farm
labour reported on the 'evils of bothy life'. One official
concern was the 'impropriety' of young men living
together and the resulting effect it would have on normal
society as it encouraged bad habits, such as drinking
alcohol. Often, the report commented, there was but one
apartment in the bothy, thus mixing living and sleeping
quarters. The official view was that the blame for the
'disgusting character of bothy life lies with the farmer.
They are aware of the unwholesome condition of them.'

It should also be remembered that bothy life was not just
for the single man. If a married man went to the feeing
market and failed to get a work contract, often he would
take work where only a bothy was provided.

Just before World War II, the County Council of Fife
put forward byelaws on 'farm bothies, chaumers and

similar premises for the accommodation of agricultural workers'. Chaumers were fairly rare in Fife, with these basic bunks situated above the stables being more commonly found in Aberdeenshire.

The Cupar Branch of the NFU and the Chamber of Agriculture both agreed this byelaw was rather onerous. Particular objection was taken to the need for immediate provision of single beds, presses and drawers. The booklet stated, 'There shall be provided a separate bedstead for each worker'.

In a comprehensive range of requirements for the bed, the booklet also advises: 'There shall be provided a clean mattress and pillow which shall be filled with straw or other suitable material; there shall be provided two blankets per worker for the period from 1st May to the 30th Sept and four blankets for workers at any other time. The blankets should weigh no more than 5lb [or 2 kilos] per pair'. Also required were lamps 'fashioned of non combustible materials' that had to be fixed to a wall, ceiling or rafter in such a manner as to obviate risk of fire.

The bothy system may have passed into history, but it has now been replaced with the provision of caravans for migrant students and harvest workers. Those farming large acreages of vegetables or soft fruit will have a number of these vehicles parked close to the farm steading. Each will house 6 or 8 workers, who will have access to communal washing and toilet facilities, often a common recreation area, too. Even these modern facilities have not escaped criticism, with some at the centre of sensational press coverage.

Cottages
In those days, my friends on the farm were the ploughmen's children who lived in the row of farm cottages.

Their accommodation was better than the bothy, and by the time I ran in and out of them with my friends in the mid-1950s they had running water and inside toilets.

Earlier in the century, however, it was very much a case of going to the shed at the bottom of the garden, or indeed into the cattle court whenever a toilet was required. Water was also taken from a common tap or a pump that could invariably be traced back to a nearby stream. It was not unusual to find the water pipe blocked, and on investigation to find a small frog wedged in the pipe. To this day, more than half the farms and former farm cottages are on private water supplies. A number of them have higher levels of nitrates or bacteria than would be allowed in the public system.

Generally, the floors of the scullery and the washroom were just flagstones laid on top of the earth though by my time the living rooms had wooden floors and were covered in carpets or linoleum. This, though, was not always the case, and many older cottages only had the flagstones throughout the building, which did nothing for hygiene or dampness in the house. This was not helped by a lack of insulation in the buildings. Many walls were just a single brick in width, a poor protection against the rigours of winter.

Recognising the poor condition of many farm cottages, the British Government put in place a rural house-building programme at the end of World War II. At that time, it was reckoned that 1 in 10 of all farm cottages were not fit for habitation. Fife Council decided that small groups of houses should be built in rural areas and although these are now in private hands, they can still be seen at locations such as Foodieash, outside Cupar, and Rossie, near Auchtermuchty. It was suggested these developments be called 'clachans', after the Gaelic word for a small hamlet.

On the farms themselves in the 1950s and 1960s improvements took place to the farm cottages as farmers realised that working conditions had to be raised if they wanted to keep the good workers. In earlier years, before these works were carried out, life in the cottages could be quite raw and damp, especially when the work outside involved cold or dirty jobs. Returning home after a wet working day that involved pulling turnips in frosty weather or ploughing in the rain was a very miserable experience indeed.

Chronic diseases associated with damp conditions such as rheumatism were the lot of the farm worker. The rural doctor had to deal with the consequences and a doctor's visit cost money right up to 1948 when the National Health Service was introduced, providing free health care for everyone. Illness or accidents were a constant concern as there was not the safety net for workers that there is now. There was no need for the farmer to pay any sick pay, although most did without compulsion, but a long-term illness or injury was a major worry for the wage earner, especially those with large families.

A long-term illness often meant expulsion from the farm cottage as the farmer would need it for an able-bodied worker. To make matters worse, in those days there was no obligation on the local authority to house homeless people.

Within the farm cottages, the work routine was equally harsh and lengthy. While wives were often working themselves, they also had the responsibility of feeding their menfolk and children, although this latter task was quickly passed down to the older children, who also helped look after their younger siblings.

By mid-century, during my youth, there was no central heating, but all the cottages had a boiler in which to wash the clothes and blankets. Many were not connected to an electricity supply until the mid-1950s, and before rural areas were connected to the national grid, many received their power from the generator on the farm.

Before electricity came along, the lighting came from paraffin lamps or even just candles. In those pre-electric power days everyone was more accustomed to going to bed early, especially in the winter months. Without electricity, the hobs on the fire were used for kettles and for cooking the meals. Sometimes a hanging chain was used for the cooking pots over the open fire.

The diet of farm workers has been described as basic, but as early as 1893 it was reported that they had given up the old diet of oatmeal, potatoes and bannocks (oatmeal cakes), and were buying meat, sometimes as much as 5/-, or 25 pence, worth every week. Vegetables from the garden supplemented basic foods such as potatoes and oatmeal, and this was especially true in the winter, with leeks, carrots and kale helping to make the base for many a plate of soup.

For any other food, there were two buying options: these were the circus of small vans, or prior to them the horse-drawn grocery and butcher-meat carts. On farms close to built-up areas, there might be bakers' and butchers' vans every other day. One housewife declared that such were the number of vehicles selling around the farms, 'Yer hands were never oot o' yer purse.'

And yet they had to be, because until the 1940s wages were paid only at the end of the first 6 months in work. Normally there would be no income until then and most households survived by local van men allowing credit to be built up until the wage was paid. This cycle of debt was then repeated on a 6-monthly basis. To go cap in hand to the farmer and ask for money to be advanced in order that the family might get through to the next payday was seen as proof of a wife's fecklessness with money.

For the working man, his morning 'piece' would usually be sandwiches with fillings of cheese or jam. Fillings such as corned beef or fish paste were also favourites. As was often said in a parody of *The Beatitudes*, when the men sat down for their morning break, 'blessed are the piece makers for they shall inherit the earth'. The pieces would normally be wrapped in a sheet of an old newspaper or some greaseproof paper. They would be put inside a tin so that the farm dog, or any vermin, would not be tempted to have an early feed at the ploughman's expense.

After World War II, every piece seemed to be carried in an ex-army khaki coloured knapsack, which often served the secondary purpose of providing a dry seat on a damp field headland during break time. Former military clothing was also used by many farm workers during the post-war years: before cabs became an integral part of the tractor, it was quite common to see the driver swathed in an army greatcoat as he tried to keep the cold or the rain, and sometimes both, at bay.

At least these coats were an improvement on the old sacks that the horsemen threw across their shoulders in wet or cold weather. Sometimes, during a heavy shower or spell of rain, the horsemen would actually take shelter under the belly of their horse.

Summer wear was invariably the bib-and-tucker set of dungarees. It took a long time for the non-farming public to adapt to denim with their wearing of jeans and they have still to adopt the bib part of the dungarees.

If they were working in wet muddy fields, the workers would invariably wear 'nicky tams'. These were created by the simple measure of tying string just below the knee, thus keeping the top half of their trousers clean and dry. Some workers, who had been in the Army, also wore bands of protective material around their legs in the manner of puttees. Today's farm workers also wear a uniform. Generally it is an overall that keeps most of the dirt and dust out; often complete with a sponsor's, or tractor manufacturer's logo. To top this off, most of today's workers wear a baseball cap, an import from the USA.

The importance of agricultural employment

Despite remaining one of the main industries in Scotland, over the past century farming has seen a slow, but consistent fall in the numbers of people employed. One hundred years ago, some 209,000 people worked the land. That was equivalent to more than 10% of the total workforce. By 2000, there were fewer than 30,000 full-time workers on Scottish farms, just over 0.5% of the total population.

The first major decline in the numbers employed on farms came during World War I. Recruitment posters with Lord Kitchener pointing out, 'Your Country Needs You', along with a heady dose of patriotism, saw the biggest-ever reduction in manpower on farms. During that period, one local Union chairman said, 'There was no doubt that agricultural men were of the best type. They had more stamina than two men. It was strong men that were needed for the war effort.'

Within a matter of four years, more than a third of all male farm workers had enlisted. And where did they end up? The answer is both simple and desperately sad: the vast majority ended up under the soil in a foreign land, as anyone who visits the vast yet neatly kept cemeteries close to the battlefields in France and Belgium can see.

To fill the gap, more women were employed and the same period saw the arrival of the Women's Land Army. The same Union chief also had a view about female labour on the farm: 'They have already proven to be very useful and I have heard very good reports of women, but they are no use for driving a pair of horse.'

After the Armistice in 1918, when the surviving troops came home, few former farm workers returned to the land. Initially, wages rose as farmers who had made big profits in the latter years of the war attempted to ensure that their farms were in full production with full staff complements. However, as imports swamped the home market and prices plunged, wages were cut and discontent emerged.

Throughout the 1920s and 1930s, the numbers of farm workers continued to dwindle. This was not a profitable time in the industry and wages meant money going out of the farm. Many of the West Coast farmers setting up in Fife attributed their survival in these hard times to being able to work the farm solely with members of their own families.

World War II saw another drift of farm workers from the land as they went off to fight the enemy. Meanwhile, those left behind picked up the skills needed to operate the tractors and other labour-saving machinery. This post-war surge into more and more mechanisation underlined the fact that the days of having large farm staffs were gone forever. However, there were still some shortages and the Union felt obliged to try and recruit workers from local towns. This had mixed success. One farmer reckoned town dwellers thought that 'any duffer can do farm work – but that is not the case.'

More recently, Fife may be one of the few areas where employment connected with farming and food production has stabilised. The introduction of labour-intensive commercial soft fruit and vegetable enterprises has brought with it a need for large numbers of harvesting hands. These seldom go into the full-time employed category and the official agricultural employment statistics do not reflect their numbers.

Hierarchy

There was a real hierarchy of workers on the farms, but at the top of the tree was the 'grieve'. He was in direct contact with the farmer and in charge whenever the farmer was away. One farmer was heard to pronounce that if he was given good gaffers, or grieves, he could farm the half of Fife.

Many of the best of them might have been farmers themselves and often they knew the farm, and definitely the farm workers themselves, better than the farmer. They were men of strong opinion and the relationship between farmer and grieve was often fraught. One story with wide circulation in North Fife related around Jim Duncan, grieve at Rathillet farm and in charge of more than a dozen men.

One day, on his return from market, the farmer, Joe Harper, said that he had bought a hydraulic loader for the tractor. This would take much of the hard work out of jobs such as emptying dung from the cattle courts. Jim Duncan thought all this a bit modern and when the new piece of equipment was delivered, he hung it from the shed couples, or rafters. It apparently stayed there in pristine condition for more than a decade as the farm staff continued to empty the courts by hand graip (forks).

While he may have come off second-best in that instance, Mr Harper evened out the score by wearing a white mark on his wellies at all times when inspecting the ploughing. The mark ensured the men were ploughing at the correct depth.

More evidence of grieves being men of independent mind comes from another farmer, who recalled that his first grieve was an excellent man. There was therefore disappointment when his employee came to hand in his notice. Asked why, the grieve – frustrated by the farmer's constant interference in the working of the farm – replied that the farmer did not need a grieve, 'only a talking orraman.'

Even in their leisure time, the farm grieves seldom mixed with the other men. One recollection was of the farm men all waiting for the bus to take them to Cupar while their grieve stood a little way off. The working hierarchy went from farmer to grieve, and onto the horse-men or tractormen. These latter categories were even stricter in who was first among equals.

The first horseman always had the best pair of horse on the farm and he carried out the most prestigious work. His team would plough the rigs close to the road, where the neighbours would look over the dykes to see what was happening. He would lead the team of binders cutting the grain and his cart would be the lead in any teamwork.

His team of horses would be first out of the stable in the morning and first back at night. It was a foolish number two horseman who tried to usurp that position, but equally, the second horseman would stamp on any indis-cretion by the man responsible for the third pair of horse, and so on down the line.

This tradition was carried on until late in the last century and I recall tractors coming back home from various corners of the farm, all checking to see they were returning in the correct order. Anyone even slightly ahead of the rightful place was expected to dilly-dally a little to ensure the correct pecking order. Equally, it was not unknown that if someone broke rank coming out of the stable, then he would be sent back in to restore proper order.

And so to the orraman: for much of the century, such men were needed to carry out the many unskilled, but physically hard jobs on the farm. The *Courier* newspaper used to carry literally dozens of advertisements for those who neither wanted nor could drive horses, nor cope with tractors. They always called for 'good workers', but who would describe himself as anything other?

The orramen were those who helped graip at the potato pits, bagged off grain at the threshing mill and carried out other menial tasks on the working farm. Orramen were scarce. They were expected to do any work required of them. Often they were only in demand at busy times of the year.

Then there were the bothy loons and single men, who took the spare pair or the single horse for work such as basic harrowing in spring time, taking the milk cans to the station on a daily basis and carting in the feed for the indoor livestock.

Those working with livestock never went into the stable to get their orders. It was accepted that those working with cattle or sheep would negotiate their work with the farmer, not the farm grieve. Rather reluctantly, the shepherd or cattleman would help at harvest time or possibly even in the spring of the year, but they always did so in their own time after tending to the needs of their stock.

Work was often a family affair, with many of the wives also working year round on the farm. Apart from milking cows and feeding poultry, they were seldom given anything other than menial work. To them fell tasks such as gathering stones off the newly sown ground in the spring to avoid damage to machinery at harvest, gathering the sheaves into stooks for drying when the binders went into the crop and carrying away chaff from the threshing mill. At turnip or sugar beet thinning, the women would come towards the end of the line that was always led by the first horseman.

Women farm workers were traditionally poorly paid. At the end of World War I, they received just over half of men's wages. This rose to three-quarters by 1938, but in many jobs such as singling beet they were doing the equivalent of a man.

And women were poorly regarded as workers, too. In 1893, Mr R. Hunter Pringle, reporting for the Royal Commission on Labour stated, 'I am of the opinion that the women in Fife are very poor workers when compared with those in Berwick and Roxburgh. The bondagers in the Borders are young, strong lasses able to handle a fork or dung graip. In Fife they are slow with the hoe, easily tired and incapable of unusual exertion.'

Women were never at the end of the line during sugar beet or turnip thinning; the farm grieve would reserve that place. From this point he could observe the quality of workmanship and control the speed of the thinning team.

Apart from the regular female workers, whole families turned out at the busy seasons. It was accepted that farm workers' children missed school to help at harvest. In the potato field, children of various ages would do this back-breaking work alongside their mother. Shepherds and cattlemen with wives and families were popular as it was generally expected that 'the family' would ensure seven-day-per-week attendance on the stock.

Feeing markets

Although this is a practice long since departed, throughout the first forty years of the century, farm workers moved on an almost annual basis. In Fife the moving term was generally Martinmas, 28 November, while north and south of the county, the preferred term time was Whitsun, or 28 May. There were complaints against the Martinmas term, especially when the weather was severe and people had to move in cold and wet conditions.

For those who had taken a new job, the move entailed putting their worldly possessions on a cart supplied by their new employer. If the 'flitting' day was wet or cold, every stick of furniture, every piece of clothing and everyone involved was soaked. Making the situation worse was the fact that the move often ended up in an unheated, damp house. To compound this, the outgoing ploughman had often left it in a mess, meaning the new tenant could not move in without cleaning up first.

A country minister at the time said he always dreaded a wet term as invariably there would be deaths in the days and weeks following the move. Often, those worst affected with pneumonia were the biggest and strongest of men.

If death occurred, it was usually the Co-operative Funeral Service that carried out the final duties. Most workers were members of the 'coopie', where they could get a dividend on their purchases, but as one ploughman remarked, 'The Co-op always has the last say on what happens to you.'

The resistance to changing the moving date away from November to the better weather that might be expected in May was that moving during the Whitsun term meant that gardens would not be put in. This was an important consideration in those days when a goodly proportion of the family's food came from the cottage garden. To get round the problem, in Roxburgh, in the Scottish Borders, men fee'd would have a day off to put their garden in at the new farm.

For those on the move the formula was straightforward. As term approached, the farmer would enquire of his staff, 'Are ye biding?' If the question was not asked, then it could be taken the farmer did not want his employee to stay.

It was almost taken as a rule that men moved every term. This constant movement was described as a 'restless spirit within the workforce' by the 1893 report into labour in the agricultural sector. The report added that, 'in many cases it is alleged the wife is to blame.'

All the moving from tied cottage to tied cottage did nothing to improve the quality of the accommodation. In 1901, a building expert quoted in the *Royal Highland Agricultural Society Journal* stated that, 'Farm labourers shift a great deal and the cottage is their house for the time being. It therefore follows they have little interest in taking care of the structure.'

On the feeing day itself at the Fluthers, the traditional fairground area in Cupar, there was a wide assortment of itinerant showmen and numerous stalls. In this throng were the farmers and farm servants, the latter often holding their hands out, palms upward, in a depiction of someone willing and able to work.

One farmer interviewed said that his yardstick was, 'a good small man was always better than a big man.' His reasoning was that many bigger men suffered bad backs as a consequence of the heavy work involved in farming in those early days of the century.

If a bargain could be struck over wages and conditions, then the farmer would hand over a coin to seal the deal. Although feeing markets were long past, I recall in the 1960s my father still paying this bargain cash, or arles, to a new employee. His handing over of a £5 note may have been well beyond the 1s fee money of the 1930s, but the deal was just as effectively sealed.

For many farm servants, going to the feeing market without knowing whether they could get a 'fee' must have been unnerving. It is no coincidence that the local paper always reported army recruiting agents attending the

Cupar Fair to snap up men willing to take the Queen's shilling rather than go home empty-handed.

To try and avoid any imbalance between the supply and demand for farm workers, in 1926 the Fife Agricultural Society set up an employment register for farmers and workers. While it got off to a good start, this was not an unqualified or long-term success. In 1934, only 17 farmers registered and 67 'servants' – as they were called in those days, where the year-long contract defined the worker as 'servant'. Fees that year were £70–£75 per annum for grieves, with foremen getting up to £70 and others, as the NFUS minutes record, being paid £60–£65.

Once the fee was taken, it was understood to be effective for the year. In 1941, a court case was heard in Cupar over an employee who had deserted the farm to which he was fee-d. A neighbouring farmer accused of 'harbouring' this individual stated in evidence that the house that the man was offered was uninhabitable. The Sheriff dealing with the case decided the employee had broken his contract, but as no money had changed hands, there was no case to answer.

The last feeing market in North-East Fife was held in Cupar in 1939. During the war years, there was a Standstill Order in place preventing farm workers from moving from farm to farm unless there was mutual agreement. Such a move also required the sanction of the War Agricultural Executive. These restrictions were eased in 1950 so that farm workers could take up employment outside the industry.

Perks
These days, almost everyone has perks, or more fully perquisites, with their jobs. If you are a city banker or chief executive of a major company, then it seems likely that

you will be given a few million pounds' worth of share options. In my previous employment, with a major newspaper publisher, staff would be offered discounts on the firm's annual publications, such as *The Dandy* and *The Broons*. Few, though, will realise that farm workers had a far more comprehensive gathering of perks than almost any other set of employees. In fact, in the early 1920s farmers reckoned that the wage packet of farm workers could be doubled if the full value of their perks was included.

John Stewart Struthers estimated the average wage of a worker was £65 per annum, but with perks this package rose to £116. He compared the reward with the pay of a carter, which at that time was only 34/- (£1.70) per week.

Potatoes, oatmeal, milk, pigs, poultry and coal all featured in the list of items that farm workers would receive as part of their remuneration and that is without mentioning the free house, or as it has been more commonly known, the tied house. Potatoes were the staple diet and in addition to those grown in the cottage gardens, workers would be allowed up to 1 ton per annum for their families.

Today, giving a family 1 ton of potatoes each year would end up with a large percentage unused, but that was a typical agreement between farmer and worker on taking the fee. In some cases, the potato perk was measured in drills, leaving the worker and his family to lift the potatoes as part of the deal. One contract at the turn of the last century specified 1,600 yards, or 1,500 metres, of potato drill.

Oatmeal was always part of the deal and a number of working millers delivered around the farms. This basic ingredient of porridge and many other meals was essential, and the general amount was round about 5 stones of oatmeal, or 30 kilos per month.

Milk would also be provided and again there was a generous quantity with 4 pints, or 1.7 litres per family per day being a common figure. Where farms did not have their own milking cows, the local milkman supplied the milk and the farmer paid the bill.

On many farms, workers were allowed to keep 2 pigs per year. The common practice was one at a time so that there was pig meat to eat throughout the year. Pigs in their own crays (or sties) at the end of the cottage gardens were fattened through food waste from the cottage, or food from the farm.

The killing was done on the farm and the local butcher would take part of the carcase as his fee. Often the pig was shared out with cottage neighbours in an informal scheme that would see the move reciprocated when the next pig was slaughtered. After the killing, the pig was immersed in a tub of boiling water, when its hair was taken off with blunt knives. It was not skinned, but hung up and gutted. The hams were then cured by immersion in brine and hung from a hook in the living room. Many such hooks survive to this day and examples of this practice are still carried out on the Continent.

Sometimes, instead of keeping pigs, the farmer would agree to his worker having up to a dozen hens. Again the perk, this time in eggs and even occasionally as chicken meat, helped supply the family table.

On some farms, where wood was scarce for fires and boilers, and these were the only sources of heat, often there was an agreement to supply so many bags or even tons of coal during the year for the worker. There was no electricity in the cottages, but paraffin for the Tilley lamps was also a regular perk of working on a farm.

In addition to these standard perks, others crept in during times of hardship. In the East Neuk of Fife, with its fishing villages, many farmers would buy a barrel of salt herring, which they left in the stable during winter. Workers could help themselves to this source of food, if they wished. In 1918, the farm books at Drumrack Farm show the purchase of a barrel of salt herring for 30/- (£1.50).

Although it was not universal practice, some lairds were also noted for their generosity at Christmas time, giving out gifts that included coals for heat and rabbits for food.

Possibly the most contentious perk of all was the tied house. Back in the early years of the last century it was a fairly common practice to have houses for employees. Many coal mine and factory owners built houses next to the workplace to ensure an on-hand workforce. On farms, the position was similar in that transport was such that living anywhere other than on the farm was not feasible. Cottages were part of the deal with the landlord on tenanted farms, along with the farm steadings and the farmhouse.

The tied house strings have now been loosened with the reduction in the tenanted sector and with increased mobility of the workforce. Very few of the rows of farm cottages in the countryside now house farm workers. Most tractor drivers and others now working on farms own their house or rent them from housing associations.

Working hours and unions

For children, Christmas morning can be a wondrous experience. I was no exception, scrambling down to the bottom of my bed to see what Santa had brought. Outside, I could hear the sound of the cattle as they were being fed and the tractors as they moved about the yard. Right up until the mid-1950s Christmas Day was a normal working day on the farm.

New Year's Day was different and always had been so.
Even when they had few other days off, this was one of the
accepted holidays for farm workers.

For the first four decades of the century, the normal
working week was 6 full days, with only the Sunday off.
Even then, there was an expectation that horsemen would
tend to their beasts on the Sabbath.

Just after World War I, the local branch of the NFU
argued against men having a half-day off on a Saturday.
One member fulminated this was the equivalent of giving
them another 26 days' holiday in the year. When added to
the 3 statutory holidays to which they were entitled, then
they would be off work for 29 days each year.

Note the 3 – yes, just 3 – statutory days off. These were
New Year's Day, feeing market day and Fife Show day.
Other than those, the only break from full-time work was
Sunday. A year or two later, an application was made to
the Fife District Council by the recently formed Scottish
Farm Servants Union (SFSU) for a sports day and gala on
the first Saturday in July. This was to be held in Thornton,
with farm workers from all over the country invited.
Again, the NFU saw this as a retrograde step. One member
pointed out that they could well be haymaking then.
Another suggested the workers give up one of their
statutory days if they wanted this holiday.

In 1927, there was a court case when farm servants had
refused to work on the fifth Saturday afternoon during the
grain harvest. The workers said they were contracted to do
4 full Saturdays, not 5, but the Sheriff found in favour of
the farmers, who were awarded damages.

With around 10% of the population involved in agriculture, it was no surprise that the Trade Union movement
saw farm workers as potential members. In 1912, they
appointed Joseph Duncan, the son of a farm worker, to set

up the Farm Servants Union (FSU) in Scotland. The first branch was established in his native Aberdeenshire. For the next few years, during World War I, his work was hampered by the increased wages earned as the nation pushed for more food. A bigger hurdle was that no sooner had a branch been established than the men were off to another area as their fee, or term of work on the farm, was finished.

The big breakthrough in union activity came in 1918–20, just after the end of the war. Farm prices slumped and farmers reduced the wages paid to their men. By 1921, wages were being reduced by up to 30%. Reacting to this, Duncan said he was not prepared to negotiate wages on the basis of price for produce; negotiations had to be carried out on the cost of living. Farmers, he claimed, had made big profits in the latter years of the war and had not paid the ploughmen a fair percentage of that money.

He concentrated his efforts on the feeing markets, urging the men to refuse anything below the agreed rate. It may have been uphill, but within two years in the early 1920s, there were some 200 branches of the Scottish Farm Servants Union.

Incidentally, the use of the word 'servants' in the title of their trade union did no more than reflect the attitudes and words prevalent in those days. In another example, the chairman of Cupar NFU in 1925 referred to the good relationship that existed between 'masters and men.'

The 1893 report by Hunter Pringle into farm workers states that, 'Farm labourers do not indulge in strikes. They either grin and bear it, or they leave the land.' This was almost right: apart from a small strike in East Lothian, there has never been any withholding of labour on Scottish farms.

A minimum wage was set by the Corn Production Act of 1917, with a base of 25/- (£1.25) per week, but this was lost when the Act was repealed in 1922. An Agricultural Wages Board was set up in 1937 and that year it agreed wages of 40/- (£2.00) for a 58-hour week for a byreman and 36/- (£1.80) for a 56-hour week for a cattleman.

It also stipulated that horsemen were expected to work 5 hours per week tending their animals outside contracted hours. With the introduction of the tractor, this became a bone of contention as some employers thought the same 'extra hours' should be carried out by tractormen looking after their machines. As it was, many came along to the farm on a Saturday afternoon or Sunday to wash and maintain their machine. No sooner had the Wages Board come into being, however, than it was put into abeyance by World War II.

It was not until 1949 that the Scottish Agricultural Wages Board was set up. To this day, it sets minimum wages and conditions for farm workers. Now the last remaining wages board in Britain, in recent years there have been several campaigns to rid the industry of it. Those who no longer wish to retain the board say that its work has been overtaken by the introduction of a National Minimum Wage, making much of SAWB's work redundant.

The farming industry is one of the few where working hours change with the seasons. In the early days, reduced hours in winter were dictated by the hours of daylight, but reduced winter hours also reflect the quieter time on the farm, with no crops to be sown or harvested. Winter is defined by the SAWB as being from the 'first Monday after the second Sunday of November and lasting to the first Monday after the second Sunday in February.'

An example of post-war working hours came with the 1948 SAWB proposals. In summer time from 7 a.m. to 12 noon and from 1 p.m. to 5 p.m., with a break of 20 minutes for 'piece time' per weekday, and from 7 a.m.

to 12 noon on Saturday. In the winter period, the hours were to be 7.30 a.m. to 12 noon and 1 p.m. to 4.40 p.m., with only 13-minute 'piece time' breaks.

The issue of mid-morning and mid-afternoon breaks, where the men would draw breath, smoke a cigarette or just sit down and eat their 'pieces' was another bone of contention for the NFU. In 1918, J. Clements of Balkaithly, Dunino, complained that if all the morning and afternoon tea-breaks were taken out of the working day, then 'farm servants only worked eight and a half hours per day.' However, a colleague – Mr Fleming of Renniehill – was in agreement with a number of farms that were taking 2 hours at lunch time instead of 1½ hours because it gave the horses more time to rest.

Throughout the first sixty years of the past century, there was a running battle between the authorities and farmers over the employment of younger people on the farm. In 1920, Fife Education Authority gave permission for potato picking, but children had to be over 12 years old. The vote was not unanimous as the chair pointed out that in Angus and Perth youngsters of 8 and 10 years old were employed. Then at frequent intervals in these decades, and especially after harvest or potato picking time, schoolteachers and education authorities publicly complained that country children were falling asleep as a result of their out-of-school-hours' duties.

Even with the introduction of the tractor to farm life problems ensued over employing youngsters on farms. Under the title of 'Vicarious Criminal Responsibility', Cupar NFU discussed the safety regulations that prevented anyone under the age of 13 driving a tractor on a farm. Union members made comparison with those of a similar age who could lead a horse around the farm without any problem.

Nowadays farmyards, with their huge tractors and massive machines, are no place for anyone other than farm workers. Those with skills are in high demand; they work extraordinarily long hours during sowing and harvest. Often, they operate machinery worth more than £100,000, and they use sophisticated electronic gadgetry that would leave their forebears gasping. With few on the ground, their output is also one hundred times that of their forefathers.

Chapter 5

Fertilisers

IT was safe in the feeding troughs; they were wide and deep enough for a small body to be far from any reversing horse and cart or tractor, and bogie or trailer. The trough was also a safe refuge from the men working away at filling the carts and emptying the cattle courts. From this vantage point, you could see the skill in peeling off layers of dung rather than delving the graip deep into the heap. You could also see the team of men working around the cart, gradually filling it, but you were not safe or free from the strong ammonia of fresh farmyard manure. It is a smell that makes townfolk wrinkle up their noses and express disgust, but the countryman views it as natural. Although it was definitely not the case, most memories of emptying cattle courts recall frosty winter days, where the heat of the dung in the courts, the breath of the horse standing waiting for their carts to be filled and the men hand graiping onto the cart all combined. The result was plumes of steam that seemed to little eyes to be mist swirling around.

The acceptance of the smell of farmyard manure may also be based on the fact that this by-product of keeping livestock was the main source of fertility in the early days of the century. The full carts would go out of the courts to a midden, where they would be couped, or tipped up, and other men – or, in later times, tractors with front loaders –

would help consolidate the dung heap or midden. Such was the importance of farmyard manure to the prosperity of a farm that it was said that farmers on a Sunday tour of the district would doff their hats to any well-made midden. There, in the midden, the farmyard manure would decompose over the summer months before being taken out to the fields in the winter.

To ensure the dung was spread evenly over the field, it was marked out with a shallow plough, or dung tam. The careful farmer would then cross these marks with others at right angles to leave a patchwork with 5-yard squares; the cart would move down one of the lines and at each inter-section, a man at the back of it would pull out a heap of dung using a hauk, as the long-handled graip with tines at right angles to the shaft was called. Completing a heavily labour intensive job, these heaps of dung were then spread around the squares by teams of women so that the next year's fertility was applied evenly over the field.

Farmers in the East Neuk of Fife and in other parts of Scotland close to the sea also used to go to the shores following any storm tides to collect cartloads of seaweed, another well known fertiliser, especially of the potato crop. One unnamed farmer with a fierce reputation decided seaweed collection was the order of the day and instructed his men with the curt command, 'Those of you wi' coats go to the shore and get wrack [seaweed] and those wi' nae coats or leggings, just go wi' them!'

The sea also provided more fertility. In 1915, Henry Watson, Drumrack Farm, Anstruther bought rotten fish from a shipwrecked boat, the *Glenravel*. The farm staff could not have enjoyed the spreading of the decomposing cargo. A few years later, in 1920, the local newspaper reported a trainload of sprats being used as fertiliser on farms in North-East Fife. The wagons were open and on arrival at the rural stations, the sprats were shovelled onto farm carts. On the roads from the station, trails of seagulls followed the farm carts. By the time the sprats were spread on the ground for fertiliser, according to the report, there was 'intensity of gulls'. The men doing the spreading were like 'spotted dicks' as a result of the seagulls dive-bombing them.

Shell fishing has played a major role in the East Neuk and often farmers would buy crab or mussel shells from the local fishmongers for fertiliser. These waste products of the fishing industry were valued for their lime content, and to this day, mussel shells can be turned up in the farmlands close to the fishing towns.

In the early days of the twentieth century when coal was a major source of fuel in the towns, chimney sweeps could sell soot to local farmers as it was also considered first-class fertiliser.

The farmer wanting to add more fertility than his own livestock could produce would buy Peruvian guano, the major fertiliser in the early years of last century. Thousands of tonnes of these sea-bird droppings were shipped across from the Pacific coast of Peru. The trade was based on slave labour bagging up the highly nitrogenous fertiliser accumulated over centuries from roosting points along the sea cliffs and then carrying the sacks down to waiting ships. This continued until well into the twentieth century.

While crops require a large number of elements in their growth, there are four main ones that affect fertility: nitrogen, phosphorus, potassium and calcium. Two hundred years have passed since farmers learned these vital pieces of crop husbandry, but even in the late 1920s, the educationalists were still promoting the need for good fertiliser usage. In 1928, a meeting was held in the Corn Exchange,

Cupar to promote nitrogenous fertiliser and the latest technology was available to ensure farmers got the message.
The reporter noted, 'The cinematograph was operated
from a van containing a dynamo from which current was
conveyed via a cable to the cinematograph machine.'

Initially crop growers would use 'straight' compounds
that contained the four vital elements. In the late nineteenth century, the North of Fife Farmers Supply Association (NFFSA) was buying fertilisers with these basic
elements. The *Transactions* of the Association were written
in copperplate and show precisely how much of each fertiliser was traded. As a provider of nitrogen, nitrate of soda
was bought from Messrs Cunningham, Edinburgh for
£12 12/6, or £12.63 per tonne delivered to the required
local station. Among the dozen farmers buying this nitrogenous fertiliser was David Berwick of Collairnie, who
bought 2 tons, 1 cwt, 2 qrs and 11 lb, or 1.88 tonnes. His
neighbour, John Bell at Glenduckie, only purchased 1 ton,
12 cwts and 15 lb, or 1.51 tonnes. Note the precision in
the billing with every pound accounted for.

Other pages show purchases of super-phosphates made
by treating mineral phosphate with sulphuric acid. This
manure was cheaper and the farmers seemed to buy it by
the ton. Bone meal was also bought. This time the supplier
was a merchant from Montrose at a cost per ton of £8
15/- (£8.75), delivered by train. As the name suggests, it
comes from grinding down bones after the fat and gelatine
have been extracted. There was also a large trade in
dissolved bones delivered in bags from a merchant in
Edinburgh to the local railway station.

The potash element came from Leopoldshalt Kainit,
which as the name suggests was imported from Germany.
Farmers requiring this manure were required either to
uplift it from Tayport harbour following its importation by

Messrs Hutchison, Kirkcaldy, or to have it delivered to their railway station. The price per ton was 50/- (£2.50) off the boat or 54/- (£2.70) from the railway vans. Although the direct trade in potash had long been finished, it was a shortage of potash from Germany during World War II that impacted most severely on crop production.

Later, as it was shown that different crops had different demands, these straight compounds were then mixed by hand on the loft floor. Those who have carried out this task recall several features, but all mention the ammonia smell that caught the throat, especially when one was breathing heavily from the hard physical work. After mixing, the bone meal sprinkled over the heap as this light, fine powder helped stop the mix coagulating into large un-spreadable lumps.

Prior to mechanisation, this mixture was spread onto the land by hand using a canvas sheet harnessed to the front of the human spreader. Working methodically up and down the fields, the spreader would take handfuls of fertiliser and scatter it on either side. The spreading method was similar to that of sowing crops right back into Biblical times. Men took a pride in being able to apply fertiliser evenly over crops in widths up to 6 yards apart. To this day any farmer or farm worker worth his salt can tell a badly fertilised field by the telltale strips. The spreading sheets were kept full by loons, or boys with buckets working from the sacks dropped off in the field.

Those who have spread fertiliser this way testify to the fact that it was both a hard and unpleasant job. The handling of acidic fertiliser soon made it clear where the cuts and hacks in the spreader's hands were located. One cure was to find spiders' webs and twist some of the web into the cut. Cobwebs were also used to help control lesions on milking cows' teats. To add to the agony, on a windy day in spring the dust also blew into the spreader's eyes as he scattered the fertiliser.

Lime spreading was also bedevilled by wind, with the workers covered in a fine grey dust in even the slightest breeze. But recognising the importance of lime to the land, the British government heavily subsidised its application throughout the time when production of food was important. At one point in the 1950s, some 70% of the cost of lime was being met by government subsidy.

The post-war years saw the arrival of compound fertilisers, which as the name suggests, combined the main ingredients. The idea was that farmers could then buy the most appropriate fertiliser for their crops. Compounds with high levels of potash were needed for the potato crop; those with high percentages of nitrogen were preferred for grass and cereals.

In the early days of artificial fertilisers many farmers judged the fertiliser by its smell, going on the theory that the more it smelt like well-rotted dung, the better it would be. They soon got over the smelling phase and concentrated on the quality of the fertiliser. In the early days it was not unusual for the heavy-duty hessian sacks full of fertiliser to go solid. Sledgehammers and strong words were then used to break the fertiliser back down into the little pellets suitable for the spreading machinery.

After use, the fertiliser bags were washed ready for re-use and on a windy May day the fences near the horse trough were used as washing lines. Nowadays, fertiliser comes in 1-tonne sacks and is transported from lorry to shed and then to field by forklift, untouched by a labouring hand.

Chapter 6

Cereals

DEEP in every countryman's bones there must be a mechanism that jumps onto alert as harvest approaches. Even in my youth, I would feel the rising tension around me. My father would go on a daily round of the fields nearest ripeness, taking an ear of grain here and another there, then rubbing them so that the individual peas lay in his hand. Next was the all-important biting on the grain to check for ripeness. Although we just about died of convulsions, my brothers and I dared not laugh when on one occasion this non-scientific testing caused a broken tooth.

Meanwhile, the men were busy hauling the old binders from the sheds, dusting them down and removing evidence of where the hens had perched and the mice had hidden over the previous months. The canvas sheets that carried the grain would be fitted and every grease point and the oil well attended to.

And all the time a watchful eye would be kept on the actions of the neighbouring farms, just to ensure they were not jumping the gun and starting harvest. If they started before you did, then they had obviously commenced harvest too soon. But if you started before them, then that was alright. When the decision to start was taken, and it was never taken until mid-morning after the early dampness had evaporated off the crop, the procession would wend its way to the first field.

The initial entry for this tractor-drawn machine that cleverly cut, then tied together bunches of wheat, oats or barley before spitting them out had been made. Men with scythes would carefully cut a strip 2 yards wide around the field to avoid too much loss of crop under tractor wheels. Thus, the first sheaves were not only hand harvested but also tied using straw. In pre-binder days, it was reckoned that a good scytheman could cut 1.5 to 2 acres of oats per day but this area was reduced if the grain had been battered to the ground by rain or vermin. The binder had a man on the back checking that the reciprocating cutter bar did not block and that there was no blockage of the revolving canvas sheets carrying the cut grain to the great invention, the knotter.

When it worked, this ingenious invention wrapped a thin piece of twine round the bunch of grain and then with a dismissive rotation of its fingers threw the sheaf out onto the ground. When the knotting mechanism did not work, and all that emerged were heaps of untied straw, there was always a poking into its innards accompanied by much muttering. My youthful joy at the start of harvest did not relate to the language used during the sorting of the various parts of the binder, although I do remember the words were often, as they say, colourful.

Sheaves were lying on the ground and the grain not yet fully ripe, so they were built into stooks to complete the ripening. The technique for stooking required picking a sheaf up in each hand and then putting one under each arm; the next move would be to clamp the two down together, intending to semi-bind the tops of the sheaves together, while at the same time ensuring the butt ends of the sheaves were correct. Five pairs of sheaves made a stook, which were always built facing due south so that the drying sun would get to both sides. If the crop was thin, it was not unknown for farmers to reduce the number of sheaves to only four pairs, as nosey neighbours could always see a sparse number of stooks, but they would not always notice the number of sheaves had been reduced. Stooking was not seen as a highly skilled job or one where great physical effort was needed. Often it was left to the women folk, the orramen and the loons to carry out the work.

The favourite crop to handle was wheat, although it was always reckoned to be 'hard on the hands,' but barley easily took the award for most disliked because the sticky awns would lodge in the sleeves and jackets. A close second prize in those irritation stakes were the sheaves containing thistles in the days before chemical weed control.

More annoyance was caused when the farm staff had to re-build the stooks blown over in a harvest gale. Often this would happen on a damp morning, thus producing a memorable combination of damp, soggy and sticky clothing. In extreme wet weather, temporary fences were erected in fields so that sheaves could be laid against them to dry.

There were other concerns when the sun was up. Stooking could be a hot, dusty job and one of the most anticipated perks would be the tin can holding the 'oatmeal'. The simple recipe for this was plain oatmeal added to cold water. It slaked the thirst for the workers, who then left the boys to slurp the soggy, cold oatmeal.

After about ten days in stooks, when the grain had ripened and the 'heat' gone out of the straw, 'leading in' would start. This was a co-ordinated operation, with teams hand-forking sheaves onto carts. These were especially adapted by the removal of the normal sides and their replacement with straw 'flakes' giving a bigger area on which to build the sheaves. The carter always took pride in ensuring his building would not only look good, but would ensure there was no sloughing off, or loss of sheaves on the bumpy farm tracks leading to the stackyard.

Integral to the leading in the field was the boy, who would drive the tractor or lead the horse between the stooks – a job which required a gentle foot on the clutch pedal or gee-up of the horse. Any jerky start or stop would encourage an oath or two from the cart builder as he tried to keep his balance atop a cart full of sheaves.

For those in the field, the best and most remembered part of the day's work was going home on top of the last load of the day. With a soft bed of straw below and the setting sun sinking over the horizon, it is part of harvest life that is deeply etched on the mind. Muscles were tired, but the satisfaction of the work done triumphed over the weariness.

The stack yard was always reckoned to show the wealth of the farm and the skill of its workforce. Thus it was the pride and joy of both employee and employer. A stackyard full of well-built stacks would be witnessed and admired all around the district. Weekends would see a gentle tour around the neighbourhood by men on their bikes and by the farmers in their cars. The skill of building stacks was one of the highest in the farm worker's list. Curiously the advice given to stack builders was that passed on to those making roads: 'Keep it braw and fou [full] and weel [well] rounded in the middle and the sides weel redd [tidy] and there's nae [no] fear o' ye'. The man with a reputation for building good stacks could always ensure an excellent fee from farmers, who not only valued the appearance of a good stackyard, but knew it helped ensure good-quality grain when the mill came around.

Sheaves were forked from the carts into the reach of the stack builder. It sounds simple and easy, but the sheaf had to be presented the correct way round and at just the right distance from the stack builder. He worked in a clockwise direction around the stack, with every layer called a 'gang'. Any slackness in delivering the sheaves was rewarded with a torrent of abuse directed at the luckless forker, or craw, as the deliverer of sheaves was called. Thus, I learned a language to this day never repeated in school.

For the connoisseur, the stacks in Fife were built differently from those north of the Tay. To understand the difference it is necessary to know that the moving binder did not leave the finished article with a square butt end. There was a long end formed at the start of a sheaf; this unevenness helped with stooking and stability as the long ends were kept to the outside.

The stooking practice was common throughout Scotland, but when stacks were built, Fifers preferred keeping the long ends to the top at all times while Angus farmers took the opposite view, with the shorter ends uppermost. In both cases, the aim was to prevent rain entering the centre of the stack, thus spoiling the grain.

North and south of the Tay, topping out of a stack was common, with the topmost sheaf turned upside down and then folded over. Unlike other parts of the country, there seemed to be little ceremony about topping out a stack, perhaps because stacks were more commonplace in Fife.

For stacks that were to be kept through the winter, the next operation was thatching, which was intended to keep both grain and straw dry. Normally, long-stemmed wheat straw was used, but some farms on the coast used reeds. In both cases, the intent was to shed rain off the stack. In some East Neuk farms, old sails were used as short-term covers for stacks to be threshed before Christmas. To hold the thatch in place, ropes made from imported esparto grass were used. These 'sparty' ropes had no great pulling strength but were eminently suitable for thatching. And that is how the grain was stored until such a date as the farmer, or sometimes the banker, made a decision.

Threshing

Some 200 years ago, it was reckoned there were more than 300 threshing mills in Fife. It was claimed there was at least one in every parish. They were installed in the period when the price of grain was good and money was invested in agriculture. Where there was an adequate flow of water, these early models used it as a power source, but for the majority, power came from horses or cattle harnessed to a central drive shaft walking round and round. There are still a number of these horsemill hexagonal shaped buildings in farm steadings in the area.

The next development was the travelling mill, which was powered by steam. One of the first travelling mills was paraded through Cupar in 1851, on its way to a number of smaller scale farms that could not afford their own mill. The framework of the travelling mills was wood and often they were painted salmon pink with red post-office coloured frames. The travelling mill was a specialist two-man operation, moving from farm to farm, with the mill pulled by a steam engine. During transit, these vehicles also had a small wooden bothy linked to the back of the mill.

Most mill men did not go home in the week and they ate and slept in their temporary home.

Often this forerunner of the modern-day road train was further lengthened with the addition of a straw baler. The weight of the mill, baler and caravan could total 8 tons, or 7.2 tonnes. Such was the damage done by this heavy machinery to primitively tarred roads in the early days of the century that proposals were made to ban the vehicles.

The National Farmers Union and the Scottish Chamber

of Agriculture both took up the cudgels on behalf of travelling mills, arguing they were essential to the farming economy. Seeking a compromise, local authorities then suggested that the vehicles could move early in the day or late in the evening – both times were chosen because tar on the road would be harder than under the midday sun when it could be torn up by the large iron wheels. As it was, the mills often moved after working a full day, reaching the next farm for a 5 a.m. start the following morning.

All this debate was forgotten when the threat of war and the need to produce food overcame the niceties of road repair. There were also issues over who was allowed to drive. One tale relates how the driver had the licence but it was the second man who had responsibility for steering. This latter task was made no easier due to the poor turning circle of steam engines.

Often the mills were owned by large scale contractors, with several units parked up out-of-season. One recollection is of a whole field of mills and balers in a field on the outskirts of Stirling.

On arrival at the farm and on setting up the machine, the first essential was to ensure the mill was completely level. This was achieved by the use of a spirit level on the main frame. Jacks and wooden blocks were placed under wheels to help the levelling process. Normally the setting up was done when the mill arrived in the evening and the first job of the day for travelling mill men was getting the engine steamed up, using coal from the farm where the threshing was to take place.

Food was also supplied from the farmhouse to the team working at the mill, although this little act of generosity came under fire in the hungry post-World War II years. Some Union members felt they should not have to feed mill men and the custom ought to be discontinued. The problem, they stated, was they were getting no extra food rations for this service.

Numbers working at a thresh could vary, but there would be one man on the stack, with another forking the sheaves precisely to the mill man, who knew exactly how important it was to feed them evenly into the big beater drums. Lumpy or uneven delivery could throw the drive belts as the drum tried to deal with a thick wedge of un-threshed sheaf. The mill man carried a special cutting knife with a slightly hooked end for the sheaves; sometimes the sheaf knife was part of a special glove. He would normally feed the sheaf head first into the drum, but if the straw had to be bunched, then the sheaf would be entered lengthwise. Feeding the mill was a dangerous job, with limbs sometimes lost in the rapidly rotating threshing drum.

Many of the mill men wore red-and-white spotted hankies around their necks to prevent grains and barley awns going down their shirts. Often they also sported eye protectors made of close mesh wire to stop grains that pinged out from the fast rotating threshing drum.

Other staff had to deal with all the output of the mill, with two of them kept busy with the straw bunches that had to be carried away to the straw stack or 'soo'. Likewise, two women would gather up the chaff in big jute sheets and take it to the cattle courts or henhouses.

The men bagging off the grain had to be hardy; they also needed to be very much aware of the main driving belt flapping away between the mill and the steam engine as such modern-day frills as guards were not deemed necessary. Health and Safety concerns were not so tightly observed in those days and horrific tales circulated of people being beheaded by these belts.

In a big day at the mill, the men bagging off the grain could handle some 200 sacks. These were not 50-kilo bags if wheat was being threshed; the sacks held 112 kilos, which sometimes then had to be carried up the loft steps to the granary. Barley was bagged in 100-kilo sacks, while oat sacks were a mere 75 kilos. The sacks consisted of heavy-duty jute and were supplied by the railway company, who would count on transporting the finished product to the mills and to the maltsters.

When the bags were full, little slides shut down the flow. The sack of grain then went onto the steelyard to be weighed and any small adjustments were made with a metal scoop and a spare basin of grain. After being tied with twine, the sack was lifted with a chain–and–ratchet barrow onto the shoulders of the man deputed to carry it away. A cheaper system was an old fork shaft and an empty barrel. Two men held either end of the shaft and tipped the full bag over it. It was then relatively easy to lift the sack between them onto the upturned barrel. Chaff could be blown directly into cattle courts or into sheets, which were then taken away for bedding. The straw came out, either into a baler or a buncher. If wheat straw was straight enough to be used for thatching or covering potato pits, it would be bunched.

Apart from the replacement of the horse by the horse-power of the tractor, harvesting of grain remained largely unchanged for the first half of the twentieth century. When horses pulled the binders, the driving mechanism came from the wheels of the binder. Thus, three horses were often needed to pull the binder. Much of the remaining work in the harvest field, such as the stooking and the leading off, remained the same until the advent of the combine harvester.

The first of the new generation of these machines

to come into Fife was delivered to Messrs Cheape, Strathtyrum, St Andrews, in 1938. It was a Massey Harris without any distinctive red company livery. Instead, the main panels were of galvanised metal. This metal attire encouraged sceptical, possibly envious neighbours to call it the 'white elephant'.

By that time, however, combines had been working in other parts of Scotland and had shown the potential for revolutionising the harvesting of cereal crops. The first of these came to Scotland in 1932. It was a trailed Clayton harvester, which was brought by Lord Balfour to his estate in East Lothian.

By 1944, in the later years of the war, Fife had eight combine harvesters, leaving the vast majority of the crop still being taken in the traditional manner with binder and threshing mill. Not everyone was overawed by the new combines: that same year, East Fife Young Farmers club held a debate about harvesting methods and those support-ing the use of a binder won. Observing one of the first combines in the county working at the harvest at Frank Roger's farm at Kenly Green, Kingsbarns, one of the main maltsters in the area described it as a 'toy'.

Merchants were also worried about the germination of combined grain because the new method of harvest did not allow for a ripening period in the stook.

In the early 1950s, a well-known maltster – Alex Bonthrone of Pitlessie – told delegates at a college confer-ence that he expected to see a swing back to the traditional binder, as there was a need for more orderly marketing of grain in post-harvest months. 'I am sure we shall next year see the man with a stackyard coming back into his own,' he claimed. Mr Bonthrone may have had that particular point wrong – never again was there a full stackyard in the second half of the last century. He was, however, very correct in a later comment in his speech when he said that there were always two points of view on malting barley: the farmers' and the maltsters'. This still holds true in the first years of the current century. At the same meeting, my father – John Arbuckle of Logie – stated there was still a need for binders in the oat crop as the straw from combined oats was less palatable than that from the stook and threshing mill method.

The arrival of the combine harvester did not immediately render the physical handling of grain redundant as the early models had bagging units as part of the combine. Sacks were filled on the combine and, when full, were then allowed to slide down a chute onto the ground. There, they were loaded onto trailers by men using a strong stick or pitchfork shaft. One man each side of the sack lifted it upright, then each holding an end of the stick, the bag was allowed to fall back onto the stick before a grunt from each person accompanied it being lifted onto the trailer floor. There were few objections from farm staff when bulk tanks on the combines and large silos on the farms removed much of the physical effort that accompanied cereal harvesting.

Combine harvesting and bulk grain handling brought with it the need to dry the grain. The old method of stooking and stacking allowed a natural method of drying grain so that it would keep through the winter, but the grain coming straight off the field from the combine often required to be dried. In the early years when the combined grain was in sacks, some farms and merchants made artificial floors in their sheds. Hot drying air could then seep through the grain sacks lying on metal grids on the false floor.

As bags went out of fashion and were replaced with bulk grain, special dryers came onto the scene. These brought the moisture level of the grain down prior to its storage in large silos. As always with new technology, there are pitfalls and more than a few samples of grain were well and truly roasted on their passage through the early driers. By burning up any chance of germination, the farmer reduces his selling options. The grain cannot go for malting as germination is an integral part of the process, neither can any that has been burnt in a dryer be used for seed – again, germination is vital. This reduces the market to one of animal feed, which may have been an important sector in the early years of the last century but now only plays a small role in the final destination of the grain crop.

Sowing and growing
In the early days of the last century, playing the fiddle brought no music to the fields. However, if it had been played correctly, then a good, even crop of grain would result, the fiddle in this case being a stringed bow that turned a small plate in front of a hopper carried by the sower. The principle was simple: seeds falling onto the spinning disc were thrown out onto the soil, hopefully in an even pattern.

The fiddle took over from the method of sowing grain adopted by man since biblical times. A rhythmical swing of alternating hands full of grain as the sower walked steadily up and down the field had scattered the seed for thousands of years. In fact, it was said that ploughmen could be recognised when they went to town as their hands swung from side to side rather than back and forward, but I am sure there were other, more telltale features that revealed the rural dweller.

Later in the century machinery came into its own, with seed barrows pulled first by horses and then with the removal of the shafts and the insertion of a tractor linkage

point. Nowadays the seeder combines with various cultivation equipment to create what is unimaginatively if
accurately described as a 'one-pass' seeder.

After seeding, the fields had to be cleared of any stones
that might blunt a scythe blade or bend a finger in the
cutting bar of the binder. This stone collecting was generally considered women's work and small gangs of them
would be seen gathering together piles of stones that
would later be loaded onto carts heading for a dump.

Stone collecting was always the most dispiriting of jobs.
Every year there would be another crop to collect and
on Logie Farm we would regularly lift 200 tons, or
180 tonnes, of stones off 150 acres every year. The arrival
of more and more stones was not, as was often suggested,
because they grew, but because of frost heaving them
upwards, and the point of a plough or cultivator then
lifting them to the surface. On my stone collecting stints,
the instruction from the old grieve was not to pick up any
stone smaller than my head. That was all right, but I don't
think the grieve realised that picking stones was a head-
shrinking exercise! Nowadays, there is virtually no stone
collection in cereal fields. Cultivators tend not to prise
them out of the soil, and where they are a problem heavy
rollers are used to thump them beyond the range of potential damage.

One other job relating to cereal growing, which thankfully slipped off the agenda before my time, was the
requirement to hand-hoe weeds from the growing crop.
This required some nifty work with long-handled hoes
which was essential before selective weedkillers were available. If a crop of cereals was badly infected by weeds,
drastic action was sometimes taken with a spray of
sulphuric acid. Because many of the weeds were broad-
leaved, they would generally soak up far more of this killer

chemical while the young cereal plant shoots would avoid the worst of the treatment.

Such was the significance of the arrival of selective weedkillers that the local paper reported in 1949 that Mr Adamson of Friarton Farm, Newport-on-Tay had that year sprayed half his cereal acreage with MCPA, one of the first generation of herbicides.

Chapter 7

Potatoes

THE first time I found out there was money in potatoes was when I was still in my short trouser days. It might seem a bit youthful to be trading in tatties, but if that is the case then a misunderstanding has crept into the statement.

The learning occurred in the days when trailer loads of potatoes came in from the field and reversed into the sheds. There, after the trailer tipped and emptied as much as it could onto the heap, an Irishman would empty the rest with a graip, throwing them up as high as he could to make best use of the space in the building.

In between trailer loads of potatoes and his hard work, he showed me the money that could be found inside potatoes. Looking around, he chose one that had a green end or had a split, and then with his penknife, he would cut it in two. Amazingly, there was a penny – in old money – inside. I never did see the sleight of hand that slipped the coin into the potato, but for some time later I was convinced that was where money came from. My delusion ended with my father taking physical action after coming across quite a heap of his potatoes neatly cut in two by myself in a fruitless search for the elusive cash. However, there must be money in the crop, as Fife has been one of the top growing areas in Scotland for both seed and ware – the eating version – for the past two centuries.

In World War II years, the county of Fife grew more

than one-fifth of the total potatoes in Scotland, a figure dwarfed only by its northern neighbours, Perth and Angus. Prior to the crop being stored indoors, the traditional over-wintering store was the potato pit. This was made by scaling out, or shovelling, the top layer of soil from a stubble field close to the farm steading or a road, making a trench a few inches deep, about 6 feet across and whatever length was needed. The harvested crop was tipped into this base and peaked up into a point, making a pyramid about 4 feet high.

The pitmen then covered the potatoes with wheat straw, which was put on vertically to help shed the rain from the crop. Additionally, the straw helped keep out the frost, as did the covering of earth thrown onto the straw to help keep it in place. If severe weather was on the horizon, farm workers then put another layer of earth on the pits. This was one of the worst and hardest jobs on the farm according to one employed in the task.

Having secured the crop against the frost, the workers often had a problem when it came to opening up the pits. In severe weather, and this was frequently when the price of potatoes rose, it often required a pick to chisel away the protective earth to remove it. Once a section of pit had been cleared, the grading out started. This was a miserable job if the weather was cold or wet as the workers were exposed to the elements. It was also difficult to get to the pits themselves in periods of heavy snow. This was particularly marked in 1947, when pits literally vanished under feet of snow. At this point the reason why pits were located close to the farm buildings or to an access road became clear. Almost regardless of the weather, graded-out potatoes could easily be loaded onto transport coming alongside.

During the General Strike in 1926 a number of potato growers farming close to the Fife mining villages found that their potatoes had been skilfully removed without collapsing the pit because hungry miners used pit props to hold up the straw and earth topping.

The second half of the century saw potato storage move indoors. Initially, this was to existing sheds, but it was soon realised that buildings put up to store a few turnips or to keep implements under cover were not ideally suited to the more specialised needs of potato storage. So, in the immediate post-war years, most farm steadings saw the addition of buildings designed specifically to store potatoes. At this stage, the traditional hand harping or graiping of the potatoes onto the heap was abandoned.

In one giant step in saving labour, this job was replaced by elevators that could lift the potatoes up into 3- or 4-metre high heaps, where they would be stored until ready to be sent off to market.

The next phase in storage came with the arrival of fork-lifts onto farms. With them came the opportunity to store potatoes in 10-cwt or 500-kilo wooden crates, which could then be stacked in the sheds. These crates have now been superseded with 1-tonne boxes and the entire Scottish potato crop is handled in these crates. Linked to this shift has been raising the height of the storage sheds.

The internal temperature of almost all of today's storage sheds is controlled so that previous problems encountered with over-heating, or at the opposite end, frosting, are almost totally overcome.

Any farmer who tried to bring his crop into store without letting it ripen properly, or who allowed bacterial disease to run rampant in the growing crop, knew that he faced a serious problem when the tubers heated and broke down. For neighbours, always gleeful in the face of the failings of others, this phenomenon was described graphically as the crop 'running out of the shed'. At the other end of the temperature range, a cold winter brought problems with frosted potatoes. Straw was always the insulator, but frosty draughts could catch the corners of sheds and when the tubers thawed out, there was a wonderful smelly, sloppy mess that could gum up the pre-market, sorting-out process.

Initially the grading at the pits was an elementary affair with hand-held riddles (wire-mesh screens) filled by a man with a hand graip. The person with the riddle first removed the soil and then with a flick of the wrist turned the potatoes over so that any rotten ones could be removed by hand. After that, the good potatoes were tipped into a sack. These sacks, weighing a hundredweight, or 50 kilos, were hand-lifted onto horse carts, with the horsemen then taking them to the nearest station: 1 ton in each coup cart.

The next advance in improving potatoes for sale was the arrival of a grader that riddled out the small potatoes (or chats) and also allowed badly diseased or cut tubers to be hand-picked off the tables prior to being bagged.

Early graders were only mechanised to the extent that one person turned a large handle that shook the riddles and turned the rollers on the picking table. Soon small stationary petrol engines took over this laborious job.

Today, handling the crop is in total contrast to the early years of the century. Most of the eating potatoes are taken to central packhouses where electronic eyes remove those potatoes with blemishes. More electronic wizardry is then used to weigh the potatoes before they end up in the sealed, pre-packed containers that lie on supermarket shelves.

Growing and harvesting
Up until the late 1960s and early 1970s, almost all the potato crop was picked by hand. Fields were divided up

into bits (or stents), paced out by the farm grieve and
marked out by branches of broom or sticks cut from bushes
at the fieldside. The measuring-out was done with a great
deal of advice and abuse from the pickers, checking to
make sure the grieve had given the same number of steps
to each stent. Adults were allocated a full stent and children
either doubled up to carry out the same length or took a
half-bit. There was quite an art to the measuring-out
process as a slight stumble during the marking-out could
help the slightly less able, or a sneaky, lengthier stride could
be slipped in if the picker had been abusive.

The squads were from the nearby mining villages but
several farmers with smaller acreages of potatoes relied on
the October school holiday fortnight to gather in their
crop. If the pickers came from local towns, then the farm
lorry was sent in with its livestock-carrying body on; with
a few straw bales around the perimeter of the cattle float,
the transport was provided. The transportation of pickers
was later upgraded with the use of old buses taking the
place of the farm lorry.

Picking potatoes was a hard and backbreaking job, but
one of the picker's daily perks was a boiling of potatoes.
The size of this boiling varied directly in proportion to the
humour of the farmer. Some were hard men and only a
few potatoes were allowed; others took a more lenient
approach, especially in years when it was obvious there
would be a surplus of the crop.

Up until the 1960s, squads of Irishmen also came over to
help with the lifting and they used communal bothies,
where their tykes, or mattresses, were filled with straw.
Generally the Irish squads were left to look after themselves and the onlooker would constantly hear squabbles
and arguments within the extended families as to who had
too much to do and who was dodging all the hard work.

In the war years, these numbers were augmented by soldiers or prisoners-of-war, housed in nearby camps. Even with forced labour from prisoners, the increase in potato acreage in the war years put a great deal of pressure on the availability of labour. As the acreage grew, so too did the demand for pickers and British soldiers on leave were drafted in to help, as were local authority roadmen. Another valuable source of potato picking labour came from the schools and there were constant claims by the farmers' Union for additional school holidays to gather the crop. Getting sufficient pickers was essential in ensuring the crop was safely stored before winter frosts set in.

Early in the century, recognising the importance of harvesting the crop before the weather went against them, farmers initially kept children off school to help with the picking. It may not have been in the agreement between worker and employer, but there was a tacit acceptance that the ploughman's fee, or agreement, included all family help when it was needed; that is at both grain and potato harvests. In many cases, working families themselves decided the extra money that could be earned at tattie time was a bonus and came before schoolwork, and in the early years of the century many school registers showed depleted attendance during the harvest and potato lifting.

In response to this, the Education Committee decided that if they could not defeat this practice then they would limit it to a two-week spell in October. In addition, they would limit the permission to children over the age of 12. This was in spite of opposition from some members of the NFU, who felt that country-bred youngsters could cope with potato picking at even younger ages than that. After many battles between the NFUS and the Education Authorities, the last official school potato holidays were permitted in the early 1960s.

In the days of hand lifting, one of the traditions of the potato field was that, almost regardless of who was picking the crop, it was left to the female farm staff to take the ingoing length of the drill and at the other end of the field, the final part of the drill. This latter 'bit' was always considered the most difficult because the harvester usually spewed the potatoes over the out-turning circle as well as the normal length. The horseman or tractor man responsible for the actual digger was the butt of much abuse as the speed he dug up the potatoes determined the pickers' workload.

In the early years, the potato harvesters pulled by the horse operated on a spinning mechanism, with rotating forks spinning the potatoes out of the ridge. Then, when tractors came along, the machinery equipment makers produced a lifting machine based on carrying the crop over a web, thus allowing most of the soil to fall through and leave the potatoes on the top of the lifted bed. That was the theory, but as anyone who has seen a solid potato ridge from heavy wet ground going up a web can testify, it did not always work in practice.

Similar separation problems occurred when the field was full of weeds. Willie Porter of West Scryne, Carnoustie remembers a 16-acre field of tatties being hand-dug because it was so full of weeds that the diggers could not cope. However, nothing daunted by the problems of the single-row elevator, the manufacturers then produced a two-row elevator digger that brought another bout of complaint from the hand pickers because of the volume of potatoes left behind.

Pickers had many tricks to ease the day's work. Some pulled all the haulm – the stems and leaves above ground – before the digger came along so that their picking would then be uninterrupted by throwing the shaws away. Often these self-same shaws were burned at break or piece times, especially when the colder weather came in. If there was a shaw fire going and it was piece time, the miners would hunker down, or sit on their heels, as they were accustomed to doing down the mines. Sometimes it was possible to bake a potato in the fire and many a half-baked tattie was consumed.

Picking styles differed, the most popular being to put the basket, or skull, between your legs and bend your back. Anyone crawling along on knees might have been pitied, but more often was ridiculed. Apart from his measuring-out duties the grieve would patrol the length of the field and always managed to find the odd potato hidden under a clod. He would ease it out with his boot, simultaneously throwing out an admonition to the careless picker.

Often there was a second picking over the same ground, with a set of harrows being run over the land to unearth any tubers not been picked the first time around. Pickers would pair up, with a skull carried between them, leaving a spare hand to pick up the potatoes. This harrowing of fields was generally detested as pickers often had to tidy up a neighbour's stent that had not been so cleanly picked as their own, or at least as they claimed their section to be.

In the early years, with poorer crops and single-row harvesters, the average picker gathered a ton per day but this was upped to twice that by the time 2-row elevator diggers were operating.

Having sufficient picking labour was always the limiting factor in growing potatoes. In 1949, farmers in Fife reckoned they had been asked to grow about 25,000 acres, but only had labour to pick 20,000 acres. Looking to the future, the local branch of the Union asked that Government be pressed for research to be carried out into mechanical means of gathering the potato crop. Today, with total mechanisation of the harvesting operation, the 2-row machine harvesters can lift up to 200 tonnes per day in a 12-hour spell, with hardly a single person employed other than tractor drivers.

Ironically enough, the breakthrough in machine lifting came in the late 1970s and 1980s, in the planting of the potato crop. By mechanically removing the stones from the ridge prior to planting, the crop now grows in a stone-free environment and when the machines with their lifting webs travel over the ground, the soil falls through to leave only the potatoes.

This is why nowadays fields being planted with potatoes heave with mechanical activity. Teams of tractors work together in an operation that will often see more than 1,000 units of mechanical horsepower preparing and planting the crop. Deep drill ploughs come first, dredging deep furrows down the field; these are followed by the 'bed tillers' and 'de-stoners' that move the stones and clods from the ridges, where the potatoes will be planted in the hollows between. The final act in this phase of growing the crop comes with the automatic planter. Originally, many of these planted 2 rows, but now the larger growers have 6-row machines.

All this is many miles of progress ahead of the labour intensive planting of the potato crop in days gone past. After the drills were drawn by a horse plough, hand planters with 'brats' or specially-cut jute sacks that hung around the human planter's head went down each drill dropping the potatoes the requisite distance apart. Always hanging a little behind the planting team was the grieve to ensure big spaces were not left between the planted tubers, as was the wont of poorer planters. The brats allowed the planter to carry up to 28 lb of potatoes at a time so this was not a job for the weak or faint-hearted. Men would work with baskets between the planting squad, keeping the brats full. The sacks full of seed potatoes had been laid out earlier and if done correctly, potatoes were always close at hand. If the sums had not been done properly, then there would be a great deal of extra work carrying the seed from far-distant bags.

Some growers adopted the system of starting the potatoes growing in trays before planting. These chitting trays made of wood could hold about a third of a hundred-weight, or 15 kilos, and were carried down the drills by two planters working alongside each other. Following on behind the planted crops came the horse with a dividing plough that covered the potatoes before the crows found out that tasty potato samples had been left out in the open. This was a tricky operation as it involved the horse walking on top of the ridge that was to be split. Later, with the arrival of tractors, the same splitting of the ridge before the tractor took place and working with the front coverers was still a job that required a great deal of skill.

Prior to selective weedkillers coming along in the 1960s, the next operation was the handweeding of the crop using hoes. This sometimes had to be repeated in fields with weed problems, especially perennials such as thistles. Along with hand-hoeing, repeated inter-row grubbing and 'furring-up' (earthing-up to increase the cover of soil on the crop) were carried out to keep down weeds. Strangely enough, in the past few years these practices have had to be relearned for those growing 'organically' produced potatoes.

The one Achilles heel of the potato crop, which is one of the twelve top providers of food in the world today, is its susceptibility to blight. The South American Indians, who first cultivated potatoes in the high Andes, would never have guessed at the future economic importance of their basic foodstuff across the globe, nor would they recognise the importance of keeping blight from the crop. Even in the relatively dry east coast of Scotland, this bacterial disease – which thrives in hot, muggy conditions – can decimate the haulm of the crop and then the spores go down into the tubers. Blight was the primary reason behind the Irish potato famine in the previous century and even to this day, there is no magic cure for it.

A grape grower in the Bordeaux region unwittingly helped the British potato industry when he mixed lime and copper sulphate. His purpose was to keep thieves away from his grape harvest – they would not eat his grapes if they were sprayed – but he also inadvertently found out that the mix was very active against blight and mildew.

Half a century ago, Bordeaux mix was the main weapon against this most dangerous of potato diseases and to this day it is still used by organic growers to keep blight off their crops.

Seed production

Two hundred years ago, potato growers may not have realised why potatoes from Scotland were healthier and therefore produced bigger crops than occurred when they planted their own seed. It was some time before scientists connected the small, difficult-to-see peach potato aphid as the reason. In warmer climes, this bug moves easily between plants, sucking the juice from one potato leaf, and then if that leaf was diseased, spreading it to wherever it makes its next meal. Thankfully the aphid reckoned that Scotland was too cool to be habitable. As a result, the potatoes grown there have far less viral disease. More significantly, this has meant a profitable trade in healthy seed potatoes.

However, it was another disease altogether that persuaded the Scots to set up a standard scheme based on the health status of the crop and that was Wart Disease. As the name implies, the disease causes great unsightly carbuncles to grow on the tubers, but worse than the damage to the look of the tubers, it dramatically reduces the tonnage grown per acre. Following its arrival from Hungary in 1912 at a time when Fife was the second most important potato-growing area in Scotland, Wart Disease spread through the native crop, decimating many of the most popular varieties of the day.

The Government took action with a special Act that prevented the growing of potatoes on land infected with the disease. In Fife, some 900 farm cottage gardens were said to be contaminated and this led to problems with farmers stating that the fields were free from disease, yet they could not plant potatoes because of the restriction.

At a meeting of the Union a member reported one of his farms had been Scheduled as a result of a farm cottage garden having the disease, but stated, 'Of course all the seed I sell comes from the clear farm.' The Scheduling of a field or cottage garden came after the disease was identified as being present in the soil. Another farmer admitted he was prohibited from selling from his farm or at least 'he was supposed to be prohibited.' A third reckoned that if a cottage garden had the disease it would be better to give the worker £20 to go away and no one would know.

Farming leaders went out to check the extent of the disease and on one of these visits Frank Christie of Dairsie Mains, Cupar found a variety with blue-coloured tubers that did not seem to be affected at all. He reported his findings to the local authority and was later informed that a Mr Berner from Edzell could supply seed of this so far unnamed variety. Subsequently, the local authority bought tons of the seed at £6 per ton and distributed it to farmers for growing.

The following year, Mr Berner wanted £10 per ton for the seed and so, because it seemed to be resistant to the disease, the Council bought more tons of the still un-named variety. A third year came along, and sensing he was onto a good thing, the supplier asked for £20 per ton. A bit of negotiation followed before Mr Christie per-suaded the Council to pay £16 per ton. In a good bit of local authority business, once the Council had supplied all the farmers wanting to buy direct from them, they sold the balance for £30 per ton. Thus the variety 'Edzell Blue', with its immunity to Wart Disease, gained recognition. Since then some eighty years have passed and many other varieties now have the same immunity, but there are still a few commercial fields and many gardens growing 'Edzell Blue'.

Meanwhile, more and more potato growers put their crops up for health inspection following the voluntary Government Scheme for the inspection of Growing Crops, which started in 1918.

In 1921, the local NFU doubted whether it was possible to meet the varietal purity standards whereby a crop had to consist of 99.5% of one variety. The biggest problem was not health or even disease but sorting out the multitude of varieties that had over the years become mixed up with each other. With the support of the local authority on land just outside Cupar, plots of different varieties were planted in 1925 to help identify rogues and diseases in crops. People were trained to 'rogue' out varietal strangers, along with diseased plants suffering from the various viruses and bacteria that affect the potato crop. Such was the success of this tidying-up of varieties that English growers bought more and more seed potatoes from Scotland.

Demand peaked during World War II, with a massive 473,000 tonnes of seed potatoes leaving Scotland in the winter of 1943–44. Word soon spread abroad that Scottish seed potatoes were very healthy and consequently grew big crops. In 1933, it was reported that a considerable quantity of seed potatoes had been exported to Spain. However, the local NFU was unhappy about this export trade, feeling that if they sent seed potatoes to other countries then the home market would be adversely affected when the following year's production from the exported seed came back to this country in the form of early potatoes.

The export trade continues to this day with large tonnages heading for the Mediterranean countries. Almost one-third of all seed grown in Scotland will now end up abroad.

Linked to the success of the Scottish seed potato industry has been the ability of plant breeders to come up with new improved varieties. Initially these were individuals from the Victorian era, during which time there had been an explosion of interest in plant breeding. Fife had its own star in this field with Archibald Findlay, who started life not in agriculture, but as a portrait painter. However, from his base at Mairsland, Auchtermuchty, he soon produced a string of new varieties of potato that caught the attention of growers.

Varieties such as 'British Queen', 'Northern Star' and 'Majestic' made his name. They were outstanding in their day and grown commercially many, many years after he brought them out. Some aver that he was not a plant breeder, merely an expert selector of good potatoes in the mixtures grown in his time. Regardless of the substance to that allegation, he was a top-class salesman, who used to meet his English customers at the local station before taking them round his latest selections.

The financial highpoint in his business came when he brought forward a new variety, which was named 'Eldorado', after the Spanish view of a golden, wealthy city. He first sold a single tuber for £100 – remember, this deal was done a century ago with values only a fraction of today's. Then he sold 14 pounds, or just over 6 kilos, for £1,400. The next offer was for all the stock of this new variety, with the buyer willing to stump up a massive £200,000. He refused this big offer – a bad business move, as this most expensive potato variety then succumbed to blight the very next year. His star waned and shortly after the local paper reported that his funeral in Auchtermuchty was a very low-key affair, his coffin carried to the grave in a farm cart.

Roguing potatoes is one of the few jobs on the farm that is still intrinsically the same as in the early years of the century. On all seed-growing farms, during the main growing periods of June and July, you will see teams of roguers walking alongside each other, working through every row or drill from one side of the field to the other. They carry sacks to take the diseased plants off.

It was this slightly odd scene that attracted the curiosity of a police helicopter in 2005, as it circled the air during the time of the G8 summit. The pilot and observer were looking for terrorists or anyone acting mysteriously and surely as far as urban police and security are concerned, there is something mighty strange about people who walk slowly and methodically through potato fields. That is why a slightly surprised team of roguers were questioned as to their potential for terrorist activities.

Normally the roguers' work is inspected by temporary staff taken on by the Department of Agriculture for the summer months. Nowadays, the top student in the inspection course is awarded the Inverdovat trophy, presented by

Harry and Jane Lang of Newport on Tay in memory of their two sons, who died in a car crash not long after attending the inspection course.

It is no longer the case, but back in the 1960s all temporary staff taken on by the Department of Agriculture to inspect the potato crops had to sign the Official Secrets Act. According to the wording, this prevented anyone from passing on any secret official code word, plan, sketch, note or document – rather an impressive list for someone whose big responsibility is to check whether the fields of potatoes have any viral disease.

Another aspect of seed potato growing relates to the need to avoid any late infection from aphids while also controlling the size of the tubers. In order to achieve this artificial curtailment of its normal life cycle, the crops are sprayed with a dilute form of sulphuric acid. Specially adapted sprayers are used for this unpleasant work that effectively kills off the haulm prior to the crop being harvested. In days when Health and Safety was further down the list of priorities than it is now, the men driving these sprayers always had red faces from working with the chemical.

Marketing

No crop grown by farmers in this country is as prone to years of surplus production or shortage as the potato. Those who were growing potatoes in the mid-1970s will recall two years when prices went through the roof as a severe drought hit all the main production areas in Europe.

In those years, prices of £300 per tonne could be achieved for eating potatoes and even those picked off the grading tables would go, not as normal to the farm dump or to pig farmers for a few pounds, but to chip shops for up to £100 per tonne.

Machinery dealers could not believe their luck as potato growers raced to recycle as much money as they could. Farm buildings were quickly erected in the same period as a fear of paying excess tax persuaded farmers to upgrade their premises. In Angus, an enterprising accountant gained a big following after inventing ingenious methods to avoid paying tax. For a while, envious eyes looked north, but composure was regained when the Inland Revenue decided some of the tax avoidance schemes were more appropriately called 'tax evasion'.

Nevertheless, it was a bonanza time when the sun not only shone literally but for those growing potatoes provided an unforgettable period. Apart, that is, from the St Andrews based farmer who, towards the end of the season, still had a few tonnes to sell. Steadfastly, he refused the £300 per tonne bid he received from his merchant. Sadly the tale has an unhappy ending: as often happens at the end of a season the price just collapsed as new potatoes came onto the market. After the market collapsed, he ended up selling his potatoes for a few pounds per tonne.

In the early years, seasons of surplus produced drastic suggestions as to how the crop might be utilised. In 1917, great excitement was aroused at the possibility of sending potatoes to make potato flour with a farina-making mill mooted to be built at Monikie in Angus; this factory never materialised. Then, in the surplus year of 1929, consideration was given to the manufacture of alcohol from potatoes as happens on the Continent. In the same year, some 2,583 tons of potatoes were delivered to the sugar factory with the farmers being paid 10/- or 50 pence per ton. There, the potatoes were made into 425 tons of dried product, then sold back to farmers at £6 per ton, this transaction proving the adage that the first loss is often the best one.

In 1935, the Potato Marketing Board arrived on the scene with a remit to control the acreage and avoid massive surpluses. Its first move was to suggest to all growers the reduction of their acreage by 7.5%. At the Union meeting, it was suggested some growers might falsely return lower acreages than they had actually grown.

Although the powers of the PMB were put in abeyance during World War II, there were no reported surpluses in that period. This was largely due to increased consumption. Before the war every man, woman and child in Britain was estimated to eat half a pound, or 0.2 kilos, of potatoes per day. During the war consumption rose 60% and by 1947, potatoes were added to the ration book with an allowance of 3 lb, or 1.3 kilos, ration per week.

The following year, 1 million acres of potatoes were grown in the UK, and from that record acreage some 8 million tons produced. Today's figures show the average consumption is around 90 kilos per annum, but that comes from a much smaller acreage as yields have increased since the 1948 crop, or doubled from the 1901 figure of 7 tons per acre.

The Potato Marketing Board is no more and the only controls on the acreage grown are in the hands of the growers themselves, who decide how much they can grow and sell. It is still the single enterprise on a farm where a fortune can be made one year and all that profit lost in the next.

Chapter 8

Flax

THROUGHOUT history, politics has guided the hand of agriculture, helping to set the economic circumstances, either for expansion with the introduction of subsidies or closing of import gates, or for retraction by allowing food to flood into the country. Seldom, though, can any single crop have received a bigger deal than that afforded to flax. The Treaty of Union in 1707 linked together the two countries of Scotland and England. Part of the deal saw Scotland taking on a share of the National Debt of England. To balance that, there would be an annual grant paid for the manufacture of linen north of the border. The Equivalent, as it was called, provided the impetus for a major expansion in the growing of flax.

Just as the Highlands of Scotland are renowned for the 'bloom' of the heather, Lowland Scotland in the years following the Treaty were known for the 'blue of the lint', which is the old Scots name for flax. Not only were the fields full of the crop, but villages and towns in the areas where it was grown rattled to the clack, clack, clack of looms turning out yards and yards of linen.

The scale of the post-Treaty industry can be gauged by the official statistics for the year 1823. In that 12-month period in Fife some 8 million yards of linen were stamped by the Board of Trustees of Manufacturers before merchants handling it received the Government cash grant.

This level of production made Fife the second most important area in Scotland at the time. Admittedly it was well behind the Angus production of 22 million yards of woven linen.

Around the same period, local historian John Thomson reckoned some 1,500 acres of the crop were grown in Fife.

Some villages even took their name from the work that dominated their trade. The old name for Dairsie, a small village on the east of Cupar, is Osnaburgh, which means a brown linen cloth. In Dairsie, and many other villages, there were literally hundreds of hand-loom weavers, who wove the yarn supplied them into cloth for the bigger merchants and would await their arrival off boats from Perth and Dundee to sell yards of linen woven in their own back sheds.

Although the linen industry has long since gone, almost every town and village has an area called the Bleachfield, where the linen was laid out after washing. Street names such as Shuttlefield and Millflat also testify to the dominant industry in the towns in those days.

In the 1830s, the growing of flax, lint or linseed fell dramatically, with farmers getting increased prices for cereals under Corn Law protection. During World War I years, there was a revival of interest in the crop when supplies from Russia were cut off.

While tenant farmers saw flax as a good cash crop, landlords were never so positive. Calling it a hungry crop that drew the goodness out of the soil, they placed restrictions in the farm leases on the area that could be grown on their land holdings. In those days, many farmers not only grew the crop, but they and their families also carried out weaving.

The best flax crops were always grown after grass that had been ploughed in, or after potatoes. It was always reckoned to do best on heavier land and a large acreage grown along the rigging or Highlands of Fife. As a crop there were a number of advantages: it was easy to grow and as a fairly competitive plant, hand weeding was often unnecessary. When it did, though, Thomson relates this was done by women working: 'in close array, in a lying posture, picking every weed with great dexterity and expedition.'

In the early days, flax also had a major disadvantage in that it had to be hand-pulled, then stooked to allow the crop to dry. Taking it off to the factory was a tricky operation. Memories relate that due to the 'slipperiness' of the crop, loads had to be netted to ensure they did not slide off the lorry.

After World War I, the importance of flax decreased and many small mills, such as the one in Dura Den operated by Alex Proctor of Blairgowrie, closed down. In their place came larger units, such as the flax mill at Falkland.

The arrival of World War II once again put flax-growing back on the agenda. In the pre-war years most flax was grown in northern France and Belgium, areas captured by German armies. At the same time, demand for material to make canvas, tents, camouflage, tarpaulins and webbing surged upwards and local Agricultural Executive Committees were given target areas of production. In Fife, the wartime requirement was some 2,000 acres.

Husbandry skills lost a generation earlier were soon resurrected and flax delivered to the factories. The local contract was with the factory at Uthrogle, outside Cupar, on the site of the former Cupar race course.

Initially, with all the focus on the war effort, there were no problems, but by 1945, there were major arguments between the farmers and the processor. As reported to the local branches of the NFUS, flax growers were extremely dissatisfied with the grades intimated to them in connection with their flax contract. A special meeting had to be called with the flax factory owners. Apparently flax had previously been graded in the field, but factory owners wanted to change this. Their preference was for it to be graded after it had been through the scutching mill where the outer skin of the plant was removed to leave the long fibrous centre which is later woven into cloth.

This argument rumbled on for another meeting or two, with the local MP getting involved, but before a resolution came an answer from a different direction. In May 1945, with some 800 acres of flax in the ground, Anstruther branch of the NFU reported that the Agricultural Executive Committee was no longer making compulsory directions for the growing of flax in the area. And in December of that same year, the factory at Uthrogle closed down, with the loss of fifty jobs. That closure ended flax growing in Fife.

In the early 1990s, linseed – a close relative of flax – was grown as the European Union decided to subsidise the growing of crops producing vegetable oil. Once again, but this time only for a year or two, blue-flowered crops could be seen scattered through the countryside. The subsidy was paid on an acreage basis and one of the conditions for those receiving it was, reasonably enough, a requirement of proof of harvest. This required the removal of the seed heads by the combine, as well as evidence of combine harvester tracks through the fields. Curiously, there was no requirement for any tonnage of harvested crop to be verified.

Such are the wondrous ways of the European Union.

Chapter 9

Pea Growing

THE closure of the Cupar sugar beet factory in 1971 left arable farmers facing a number of problems. Sugar beet had been an integral part of the rotation on much of the good land in the East Neuk and the fertile Howe of Fife. It required large amounts of fertiliser; if residual sugar-beet tops were eaten by fattening sheep after beet was lifted, grain crops grown the following year did not require a great deal of manure.

In those days, continuous grain growing was not seen as good farming and a replacement crop was considered necessary. There was also the loss of the cheque from the sugar beet crop. This was always a helpful injection of money into the cash flow of the farm.

Some growers looked north and saw their neighbours in Perth and Angus growing acreages of strawberries and raspberries. In the early 1970s factories were still processing these crops in Forfar, Brechin, Montrose and Dundee. The Tay Road Bridge, opened in 1966, provided easy access to markets that had not existed previously, due to the slow, costly ferry transport over the river. It was therefore no surprise that the acreage of strawberries and raspberries grown in Fife increased in the 'post sugar beet' period. But, more significantly, the Cupar beet factory closure provided the impetus for vegetable growing in the area. Growers went south to visit pea-growing operations in

Lincolnshire, where they also saw the possibility of growing field scale brassica crops.

As a result, in 1971 Fife Pea Growers was set up as a farmer-owned co-operative with the aim of growing and marketing vining peas. This was followed in 1973 by the formation of Fife Vegetable Producers, another co-operative which started to grow and process carrots, cauliflower, turnips and a relatively little known crop called calabrese, or broccoli.

One of the main requirements for financial success in growing peas that will end up in supermarket deep-freeze cabinets is to have a long harvesting period. Only by keeping the harvesting machinery and factory processing lines going for as many weeks as possible can costs be pruned back.

Fife possesses a wide range of soils and climate within a fairly small radius. Plans were made to utilise the regions with a reputation for earliness so that the co-operative could start harvesting in late June. This left the later, heavier soils to grow crops that would not be harvested until late August.

Varieties were chosen to accentuate the length of season with early varieties being sown as soon as the ground would allow. Then, in late May, crops were sown which the harvesters would come to at the end of the harvest.

Growing peas has several benefits, one of which is that the plants are nitrogen fixing. This means they take the nitrogen component of their needs from the atmosphere, considerably reducing the fertiliser bill. Unlike sugar beet, which was generally considered a hungry feeder, the pea-crop leaves land in a more fertile state, but as many a farmer found out, one of the drawbacks of growing peas was that pigeons soon passed the word around that their favourite food was now in ready supply. Those farmers who enjoyed shooting found sport on their own doorstep, while those who did not carry a gun were left to tear their hair out almost as quickly as pigeons stripped the young seedlings.

Pea fields could be identified at a distance as they were often festooned with a range of devices aimed at preventing the field becoming the breakfast, lunch and evening meal for local pigeons. It was generally reckoned once the ground was covered in young plants, the bird threat receded, only to rise again as pods filled with sweet-tasting peas.

Crops were cut when they reached the correct level of ripeness. Testing equipment called a tenderometer was used to determine if peas were at a stage of ripeness that would get them either into a 'tender young pea' package or classified 'Bertie Bulls-eye' types, dispatched in bulk to catering outlets such as schools or prisons. The difference in financial reward between the two markets was considerable.

Once the decision was made to begin the harvest it was a 7-days-a-week, 20-hours-per-day operation. Initially, the co-operative used viners pulled by large tractors and this produced a spectacle never before or since seen in the area, with up to twelve harvesters travelling in slow convoy across the field. It was not much fun for motorists caught behind this modern-day, slow-moving caravan of vehicles winding their way around the narrow country roads.

Before these harvesters came, cutters swathed down the field, avoiding patches not yet ready or over-ripe. Cutters also avoided any wet, boggy parts, and as the harvesting group occasionally found to its cost, a 10-tonne viner stuck up to its axles could only be extricated with the help of some hefty gear. The getting-stuck syndrome became more severe after the original viners were replaced by self-propelled versions that cut and vined at the same time. These monsters weighed in at 21 tonnes.

The speed of the whole process was determined by the capacity of the processing and freezing factory. Harvested peas had to be frozen within 2 hours; this required a complex operation with viners being stood down for short periods when the factory fell behind in its processing schedule. It also necessitated a fleet of delivery lorries leaving the field on a half-hourly basis. Communication between the two parts of the operation was critical and the use of a two-way radio was vital.

When the co-operative was at its peak, some 3,600 acres of peas and beans were grown, with some 8,000 to 9,000 tonnes of frozen produce emerging. As one director of the co-operative claimed, this was 'a helluva heap of frozen peas!'

In the early years of the operation, the produce went to Union Cold Storage factory in Glenrothes, but when this closed in the early 1980s, pea and bean freezing was moved to Dundee and Edinburgh.

The work entailed using a double shift of men, mostly gathered from farmer members of the co-operative. It was also one of the last occasions where large numbers of farm workers would work together. The highlight of the night shift was the 'fish supper break' with the shift clerk despatched to the nearest chippie with an order for 25 fish suppers. Nothing was ever proven, but it is believed the clerk got his supper free from the chip shop owner delighted by this massive boost to his nightly turnover.

Another enterprising member of the group made a point of collecting all the empty lemonade bottles and whenever he had a sack load, he would offload them at the nearest shop that paid 5 pence a bottle for the empties.

By 2000, the pea-vining operation was discontinued as

price pressure from buyers made the enterprise uneconomic. Growers may have been able to accept low
incomes because of unseen benefits such as the fertiliser
boost that pea growing gave to the following crop, but as
grain prices rose, the pea enterprise faded out of existence
in Fife.

Chapter 10

Other Vegetable Crops

PRIOR to the 1970s, Fife had no great reputation as an area for growing vegetables. Edinburgh had Mussel-burgh, with its vegetable-growing holdings, while Glasgow had access to the fertile Clyde valley. Apart from a few smallholdings, there was no equivalent fruit- and vegetable-producing area in Fife. This was largely due to poor transportation links to the big cities and centres of population. Vegetables tend to perish rapidly and before the days of refrigerated transport proximity to markets was a vital factor.

However, with better road links and improved methods of keeping vegetables fresh, Fife has come into its own; the move into field-scale vegetable crops was also boosted by farmers looking for a replacement cash crop following the loss of the sugar beet factory. It may be claimed that, so far, climate change has helped the Fife vegetable industry. Within the UK, Lincolnshire used to be considered the main vegetable growing area, its flat, deep loams ideal for growing a whole range of crops, but hotter, drier summers in the South have dented that reputation. Fife, with its proximity to the cool North Sea and abundant supply of water for the irrigation of crops, has entered the vegetable-growing frame.

Census figures for 1901 show that Fife only grew 23 acres of carrots and 88 acres of cabbage and its near relation, kohlrabi. It is unlikely even that small acreage was destined for human consumption as sheep farmers often grew cabbage for the pedigree stock. Fife now has more than 4,000 acres of field-scale vegetables, including carrots, broccoli, cauliflower, Brussels sprouts, lettuce and shopping turnips.

The transformation is remarkable, but it could have been so different. In the early 1970s a group of young farmers, having observed the death throes of the sugar beet industry, decided to look at other crops. Their first choice was tulip production, partly because 'it was difficult' and unlikely to be attempted by anyone else. There was, in those days, a market for forced tulips for the Christmas trade in Glasgow. The business plan was that the group would produce tulip bulbs for the Clyde Valley glasshouse industry, which, in turn, produced the flowers for the city's festive period. Trial plots of tulips were planted and lessons learned as to soil suitability.

What was in no doubt was the excellent health of the tulip crops. As in the seed potato industry, there were little or no viral or other disease problems. Soon, the small group was growing 12 acres of certified tulip bulbs; the total acreage in the UK that year was only 13 acres. However, the venture floundered when the price of oil rocketed and the Scottish glasshouse industry collapsed under the burden of high fuel costs. The local bulb growers had lost their market, but rescue was at hand with a Lincolnshire-based company taking over the whole operation and moving it south. The Fife pioneers, in the words of one member of the group, had 'enough money left to take our wives out for tea.'

The next venture for this still-enthusiastic group of young farmers involved the growing of broccoli or calabrese. Advice about this brassica came from the

Scottish Crop Research Institute just outside Dundee. The crop was largely unknown in Britain with only a small acreage grown as a winter vegetable in Cornwall. A visit to Calabria in Italy, where again it was grown as a winter crop, convinced the Fifers they should try it. So, the first crops of calabrese were grown in Scotland, and with a basketful of the harvested heads, the sales manager was sent to Glasgow wholesale market. The disappointing result was that none of the traders recognised the crop; equally, none wanted to buy it. At this low point, one of those who had undertaken the expedition to Italy thought they might have been unduly influenced by the local wine. However, the group persevered and today it has grown into East of Scotland Growers, a farmers' co-operative that now produces up to 14,000 tonnes of broccoli annually. This makes it not only the largest grower of the crop in the UK, but the third largest in the whole of Europe.

As the years have gone past, the trend towards specialisation occurs even in the broccoli crop. The average grower now grows 240 acres, well above the initial 20–30 acre range. Currently some 90 million broccoli plants are hauled up from propagation units in the south of England to be transplanted in the arable fields of eastern Scotland. A planting programme ensures the harvest extends from June to November. From early March until late June, tractor-towed planting machines, each with three or four operators, can be seen moving slowly up and down the fields. Broccoli planted in early season will then be quickly covered over with polythene to prevent a nip from late frosts.

Irrigation of vegetables crops is now essential for all growers (most irrigation is carried out by large rain guns that are fed by an umbilical polythene pipe up to 400 metres long). This far surpasses the capability of the original irrigation units that relied on aluminium pipes fitted with sprinklers. After applying the necessary water to one section of the field, these unwieldy dripping pipes would then be hand lifted to another part. This provided the worker with a combination of water dripping from the pipes and an attempt to walk through a mud bath.

Workers christened the operation 'more irritation than irrigation' and the name-calling increased with the arrival of the Laureau tractor-mounted irrigator. At first glance, this French invention was a big improvement. Water was thrown out of a rotating boom carried on a tractor that would periodically be moved to a new site. However, the reality was that this ungainly machine's 15-metre long arms swung about during transit just as easily as when they watered the field. It was an absolute danger to life, limb and the local electricity supply. On more than one occasion, the rubber tyres of the tractor were blown off as the boom hit a power line. If that was not bad enough, the inventor had not realised that, after irrigating in a circle around itself, the tractor then had to be extricated from the muddy, sodden soil.

Today's rain gun, fed with an umbilical pipeline, irrigates as it is pulled down rows of dry soil. With its swivelling head throwing water from side to side, it can apply up to 72,000 litres per hour. Access to a plentiful supply of water is therefore essential and many farms now have reservoirs that hold the millions of gallons of water needed for their vegetable and potato crops.

In the early growing days vegetable harvesting was carried out by former miners from the villages of west Fife; when that source of labour dried up, immigrant workers came from eastern Europe and today almost all the crop is hand-harvested by Poles, Czechs, Latvians, Lithuanians and Romanians. When the crop was first grown, there was a certain amateurism in the harvesting and so, apart from the knives used to cut the crop, another essential for the harvesting gang was a large box of sticking plasters for emergency repairs to fingers.

Most harvested broccoli heads go south to the freezing factories in eastern England, with lorries travelling down overnight to get the produce into store. A percentage of the broccoli crop does go to the fresh market through a major vegetable pre-packer, Kettle Produce, based in Kingskettle, which supplies a range of vegetables to most of the major retailers in the UK. After starting life in the mid-1970s, privately owned Kettle Produce has grown into one of Fife's largest employers, with a workforce of more than 500 and an annual turnover in excess of £80 million.

Local farmers grow a range of brassicas, lettuce, carrots and swedes for the company. Due to Kettle Produce's major investment in packing technology, Fife is now one of the leading areas in the UK for growing carrots for the main retailers. In husbandry terms, the seasonality of crops grown today is being extended to supply the major retailers throughout the year. Early crops go under polythene to protect them from frost. Later crops of carrots are covered with straw so they can be harvested throughout the winter.

Initially the supermarkets' requirements were for a washed and packed product. Now, with market advances, demands have increased and changed so that mixed packs of ready-to-use vegetables form part of the order book. The emphasis of the whole operation is based on the requirements of the consumer. Produce for processing is brought into the packing station by trucks from southern Europe when they are totally out of season elsewhere.

Chapter 11

Soft Fruit

WE called them the 'berries' – 'them' being the new school blazers that appeared after the school holidays. The name came about because the owners had picked raspberries during the summer months to buy their new clothes. Berry pickers have always worked on piece-work rates, either receiving so much per pound, or per tray picked so these jackets would represent several hundred pounds of harvested raspberries.

Fife never had a large commercial soft fruit acreage – Perth and Angus were the big fruit growing areas. It was in Blairgowrie in Perthshire at the beginning of last century that growers found an area where summers were not too hot and where there were plenty of pickers to harvest the crop. From there, processing factories were set up, initially as canning and jam making enterprises, and then later joined by specialists in the freezing of fruit. In fact, it was stated that the growth of the raspberry crop in the years between the two World Wars was brought about by the popularity of the proverbial 'jammy piece'. For those unfamiliar with the spreading of raspberry jam on bread, this was a regular part of the diet for millions of school-children and factory workers. And if you did not take a jammy piece to school or work, you would likely get one for your tea when you came home.

Until the opening of the Tay Road Bridge in 1966

growers in Fife did not have access to any processing opportunities and this prevented its contribution to the soft fruit industry in its two neighbouring counties of Perth and Angus. At that time, these two areas grew three-quarters of all the commercial raspberry crops in the UK.

One of the largest fruit farms in Fife was at Gilliesfaulds, right on the outskirts of Cupar. Pickers came from all directions, many just walking or cycling from the town to the berry harvest. In the early days, families of travelling people would often encamp on a soft fruit farm and stay there for the season, with every member of the family turning out for the picking, but as picking standards rose, the travelling people melted away.

For many, travelling people or not, berry picking was a real family event. While the mother and older children were picking the fruit, younger ones played in the puddles, or the dust at the end of the dreels, or rows. Sometimes the youngest of the family would be wrapped in old fertiliser bags and left to sleep while work went on around them.

Fruit growers out in the country had to bring their squads to the farm by whatever means of transport they had. Some used old lorries with tarpaulin covers and a few straw bales for seats and went down to the mining villages in west Fife for their workforce. Latterly, former Corporation buses were used for transport, and even out of season, berry farms would be easily identified by the string of old buses in the stackyard. Driven by the men on the farm, because the law said they were not being used for 'hire or reward', anyone with a current licence could, and did, drive these buses.

It was an unlucky day for all concerned when the berry bus was stopped en route for the farm by the men from the 'dole'. Those quick enough to escape from the back of the bus would return to pick another day and any number of false names and addresses were provided, although the civil servants carrying out the raid already knew they were not interviewing John Smith, far less Mickey Mouse. For the farmer with orders to meet, such an event disrupted picking plans not only for that day, but several more to come.

A good grower always divided his picking squad into those who could pick good fruit – 'the specials' – and those who could not. Pickers who squeezed the fruit when picking were given buckets; those with a gentler touch would be given 'luggies', which could hold several punnets and were tied around the waist to release both hands for picking. In the fruit-picking hierarchy, being asked to put your fruit into punnets was a plus as pay rates were higher.

Originally, punnets were made of wood chip and would hold 1 lb (approximately 0.45 kg) of fruit; later, they were made from pre-formed recycled paper. Nowadays, all punnets used are made of polythene and weights vary according to the whim of the buyers.

Full punnets were transferred to trays and they in turn were taken to the weighing stations, where the quality was inspected and the piecework rate was paid in cash. Often the weighing stations were manned by students, some of them going on to become entrepreneurs in their own right. One bought potato crisps at the local cash and carry, and then sold them on to the pickers at a profit. On one occasion, there was an added bonus as the crisp company was offering cutlery in exchange for empty packets, so the student sold the crisps, collected the empty packets and ended up with a canteen of cutlery!

Picking conditions varied a great deal. The first pick of a crop could provide a tidy sum for the picker's pocket at the end of a day. A poorer, thinner crop, or rain coming midway through the day, thus reducing earning capacity, would mean a very discontented gang of pickers.

Today the arrival of the mobile phone provides an immediate network through which pickers in various parts of the country inform each other of pay rates and fruit quality. Growers will see their picking team quickly melt away to other farms where it is reputed picking is easier and rates are better. Even before the advent of the mobile, pickers were quick to find out who was paying more and where the berries were better. To combat this, the growers decided on recommended standard rates that were published. This started in 1937, when the suggested rate was half a penny per pound, or £0.05 per kilo for the first 3 pickings, rising to three-quarters of a penny per pound, or £0.07 per kilo, for later, or thinner pickings.

Those picking for the processing market were also paid on a piecework basis similar to those picking for the fresh market, the only difference being that a reduced rate was always paid for picking pulp. They would come to the temporary weigh stations, with their buckets of fruit being weighed before tipping the berries into large barrels.

Handlers had to be alert to the tricks of pickers who falsely increased the weight of the fruit they brought into the weighing stations. At the lower end of the crime list were those who put small stones in the bottom of their punnets and then covered them with fruit. At the unmentionable end were those who, hidden by the rows of raspberry canes, decided to relieve themselves in the buckets. Such action was reputed to contribute a little tang to the raspberry jam and it did add a little to the overall weight of the fruit for which the picker was paid. This of course occurred in the pre Health and Safety Executive days when there was no statutory requirement to provide on-field toilets.

Perhaps few jam buyers knew that after tipping fruit into wooden, or later on, polythene drums, the preservative used by the processors was sulphur dioxide. This chemical left the once-red raspberries and strawberries a ghostly and ghastly shade of white, but it did keep the fruit preserved until needed in the factory lines.

Producers who grew for the processing market were always vulnerable to imports of cheaper pulp from Eastern Europe. In the mid-1950s the price dropped from £165 per ton to £55 per ton for pulp. This was partly on account of those countries behind the Iron Curtain trying to raise hard currency, but it also reflected a change in eating habits. No longer was the jammy piece an integral part of the diet of the schoolchild, or the factory or office worker: more sophisticated processed foods were becoming available.

Because soft fruit has a short shelf life, time is very much of the essence when it is picked for the fresh market. Formerly the big cities had fruit markets that started trading at 5 a.m. and supplied the individual greengrocers. Vans of fruit picked the previous day and held overnight in a cooler did these early-morning runs to market.

For fruit heading south to the big city markets in England, the train timetable decided when it could be picked. In Cupar the overnight express took the fruit down to the large conurbations.

Nowadays, big supermarkets dictate collection times and again the picking force, which today is almost totally migrant labour, work to that schedule. Once the fruit is picked, it goes off to the packing station where punnets are inspected to ensure it is up to grade. The hierarchy of the labour force ensures that only reliable workers can work in this sorting shed. For the grower, the fickle demands of the major retailers mean a constantly shifting punnet size and pricing label as the supermarkets expect growers to carry out all the pre-sale operations.

The traditional soft fruit harvest ran for a month or so in the summer, but now with improved husbandry techniques, fruit picking stretches from May to October; harvesting is further enhanced by the use of large polythene tunnels. Supermarkets expect fruit every day and they do not consider it an excuse to say that 'rain stopped picking'.

As consumers become more aware of the health-giving properties of eating fruit, more and more large polythene tunnels are being erected to grow raspberries, strawberries, and recently, blueberries.

Chapter 12

Sugar Beet

THE intense sensation is now long gone but in the middle part of the last century, when the sugar beet factory was in full flow, residents of Cupar and visitors to the town would recognise the sweet, warm smell of sugar being distilled.

Cupar lies in a bowl and when I was a schoolboy, coming down into town on winter mornings, often there was a low-lying shroud of fog, pierced only by the steeples of the kirks and the Corn Exchange. On such mornings you could be sure that as soon as you descended into the mist, you would pick up the all-pervading, but not unpleasant smell from the factory, a mile to the east of town.

Even in those 1960s days, when road transport was only a fraction of today's levels, the school bus would often be held up by lorries, tractors and trailers heading for the only sugar beet factory in Scotland.

Before we look at how the factory came to Scotland, and how it closed down less than fifty years later, we should examine the importance of sugar in the Western world. Although today fingers are pointed at sugar as one of the underlying reasons why we have so many diabetics and generally obese people, a century ago it was called 'white gold'. It was a valuable commodity that was expensive to obtain. The importance of the Caribbean sugar-cane growing industry was one of the main reasons behind the slave trade from Africa. As an alternative to sugar from cane, a German – Andreas Marggraf – found out that it could also be extracted from a root crop, and thus sugar beet originated as a crop.

The trigger for the UK growing more sugar beet at home was undoubtedly World War I. Not only was shipping from the sugar-cane growing colonies under fire from enemy guns, but the main sugar-beet growing area on the Continent – on the French-Belgian border, to be precise – was being turned into a sea of mud and carnage as the two sides threw millions of men and tons of ammunition into trying to win the conflict. The resulting scarcity of sugar pushed the price upwards and breathed life into homebred refining. To encourage this young industry, the Baldwin government pushed through the Sugar Act of 1925, allowing subsidies for new factory building.

The scale of Government support for this new industry was evident in a 1931 House of Commons answer, which showed a total of almost £500,000 going to the company in financial guarantees. This subsidy, which helped set up the Cupar sugar beet factory, came into being as part of the Anglo-Scottish Sugar Beet Corporation. Although a small sugar beet factory was built in Greenock, it only had a short life, going into liquidation in 1928 and leaving Cupar the only beet-refining factory north of the Tyne.

In 1935, following a major Government review, the ownership of all sugar-beet factories in the UK was rationalised, and for the rest of its life the Cupar factory came under the ownership of the British Sugar Corporation.

The early years of the Cupar factory were difficult. Not only were farmers learning to grow a new crop, but the science used to extract the sugar from the beet was known to only a few experts. In 1924, two years before the factory came into operation, trial plots of the new crop were sown, both in Fife and Aberdeenshire. The results in Fife showed that crops of around 7 tons per acre could be achieved, but further north the trials produced only around a ton and a half – a yield far from producing any profit for farmer or processor. Following these trials, Cupar NFU members were addressed on the intention to build a refining factory that could cope with up to 400 tons of roots per day. The potential growers were promised beet would be paid for at 44/-, or £2.20, per ton with a 2/6d, or 12 pence, bonus for every degree over 15% sugar content and a similar level of deduction below that par.

On Christmas Day, 1925 the first sod was cut on the Prestonhall site outside Cupar, a location favoured by the Town Council as they had promised a water supply to the new venture. Less than 12 months later, on 8 November 1926, the first beet was sliced. The factory had been built at a cost of £300,000 and some 2,718 acres of sugar beet were contracted that first year. It had also been built with access to the main East Coast railway line and in the early days, more than half the delivered tonnage came in by railway wagons. This method of delivery was favoured by beet growers in the Borders, Perth and Angus although also popular in pre-World War II years, when fuel rationing curtailed transport.

In 1927, recognising that motor transport was also important in the delivery of sugar beet to the factory, Cupar NFU approached the factory management to ask if they would be prepared to purchase a number of lorries and then hire them out to farmers. This request may have been politely received, but no action was taken and right up to the closure of the factory, farmers delivered their own beet by tractor or lorry, or employed haulage contractors to do so.

The factory normally worked on a three-month 'campaign' as it was called because the beet had to be processed soon after it was lifted in the autumn. To cope with this seasonality in production, the factory owners often augmented local employees with men from the Western Isles of Scotland. For example, in 1951, 100 men from the Isle of Lewis were recruited. Among them were 30 members of the McLeod clan – 7 with the forename of Malcolm, 5 called Donald and 3 named John. There were also 12 McKenzies and a number of MacDonalds. To ease identification, the factory owners provided the Lewismen with army-style numbers. The local paper reported only 99 arrived – 1 had fallen asleep on the boat and missed the landing on the mainland at Kyle of Lochalsh. These temporary workers lived in special hostels on the factory site and worked 8-hour shifts, 7 days per week, with 1 day off every 3 weeks. If they broke the contract, they had to pay their own fare home.

In the 1960s, at the peak of operations, some 200 people were employed at the factory during the 90-day campaign. Those working on the 'campaign' would often start in mid-October and continue right up to, and occasionally beyond, the New Year. During the campaign period, the factory sent out delivery permits to growers so that they controlled the input of beet. Complaints were often made that certain farmers were favoured by this system, finishing their harvest early and thus avoiding lifting beet in early December, when the fields were muddy and lifting was difficult.

The other major problem area for farmers was the sampling system. Beet was paid for by its sugar content, which was established from a sample of beets drawn from the delivery vehicle. However, the same sample also established the level of deductions for earth, or tare. Some samples recorded more than half the weight as earth or badly topped beet. The level of tare had a massive effect on the final pay cheque; the percentage of sugar in the beet was also important in the final payment. Normally, sugar beet contained about 16% of sugar, but payment was on a sliding scale. Some samples went up over 20%, while others, possibly with 'shot' beet in them, were down in the lower teens.

In its natural state, sugar beet is an annual plant, throwing its seed head in the season in which it is planted. When it does this, the sugar content of the beet itself falls dramatically. It was essential that only beet that had not produced a seed head fell into the sample bucket.

Once through the sampling process, the lorry was emptied by a water cannon of a type that nowadays is sometimes seen in use as a means of quelling human rioters. However, this method of emptying vehicles did not come along until the late 1940s. Prior to that all beet – from lorry, trailer or railway wagon – was emptied by men with hand graips. These graips had metal buttons on the tips of the tines as beet could 'bleed' and lose some of its inherent goodness.

In order to keep the workforce busy over a longer period than just the slicing campaign and also to supply the Scottish and North of England demand for sugar, the Cupar factory also packed the refined sugar. In the early days, the owners tried to provide year-round employment by importing cane sugar. This was shipped to Dundee and then transferred by rail to Cupar. In the first decade of the factory's operation, some 68 cargoes of cane sugar were landed at Dundee. Thereafter, economics appear to have ruled out this operation.

Initially, the refined sugar went into large, 1- or 2-cwt (50- or 100-kilo) jute sacks, which would be stored 20 to

30 sacks high in a massive store before being taken away in lorries to jam-making factories and catering outlets. Latterly, a large amount went into 2 lb (0.8 kilo) retail packs, following a major investment in special packaging machinery in 1965. The handling of heavy jute sacks full of sugar came to an end in that same decade, with bulk tanker lorries taking over the delivery role to major commercial customers. These bulk vehicles were filled from the large concrete silo, which still dominates the skyline over the bare bones of the former factory. The silo could hold some 12,000 tonnes of refined sugar and it was built in 1964, less than ten years before the factory closed its gates. Today, it is still in use as a grain silo.

In another attempt to keep the factory wheels turning, in 1929 around 1,000 tons of surplus potatoes were delivered for drying into animal feed. The experiment, supported by the Cupar branch of the NFU, was not a success as it took more than 6 tons to produce 1 ton of stock feed.

Throughout its lifetime and right up to the closing of the factory doors in December 1971, the relationship between factory and farmer was fraught. In the early days, there were always discussions between the two over prices and conditions. Soon, the factory employed a Farm Liaison Officer to deal with the issues. Beyond these local concerns, there was also a constant sniping and comparison with conditions in the other sugar-beet factories in the east of England.

In 1933, when Scottish growers did not grow as much beet as was required by the factory, the secretary of the Norfolk branch of the NFU suggested the Cupar factory be dismantled and brought down to Norfolk, where it would be fully utilised. Similar comments were made the following year, when once again there was a threat of closure at Cupar because the factory had not been working at full capacity and was in fact running at a loss.

One year later, in 1935, a proposal to close the factory was met with a Union organised protest meeting in the Corn Exchange, Cupar, supported by a whole range of interested parties, such as hauliers, agricultural suppliers and tradesmen. Addressing the meeting, local MP, J. Henderson Stewart, said that the loss of the sugar-beet crop to the arable areas of Scotland would be immense.

A further wrangle came just before World War II, when it was discovered that out of the £63 million poured into the UK sugar industry in the period between 1925 and 1939, only £1 million had been used to support the Scottish sector. Indeed, throughout its short lifetime, with the possible exception of the war years, the Cupar beet factory was never far from the threat of closure.

The acreage needed to ensure it operated at full capacity was close to 15,000 acres and that area of good beet-growing ground did not exist within economic distance of Cupar. Throughout the second half of the 1960s, the acreage needed by the factory was never met and warning shots were fired over the heads of beet growers by the company, indicating that unless more acres were grown, the factory would close. The threats became more substantial as the decade came to a close and even if the acreage had risen by then, the company claimed the Cupar factory – one of the smallest in their ownership – was losing money. Frantic meetings were held with politicians, both local and national, and yet the die was cast and the last load went into the factory at 4.30 p.m. on Thursday, 23 December 1971.

Of course, all this was a far cry from the early days in the war when a local resident offered to grow a few rows of beet in her garden in Cupar, if it helped with the overall tonnage going into the factory. A tactful response came with the Liaison Officer asking that she co-operate with a few neighbours to get the area up to at least half an acre, at which point they might be considered a registered grower.

Cropping

Sugar-beet singling has been described as the 'worst of jobs'. The word 'singling' is distinct from thinning turnips; it entails not only the removal of all weeds from around the small plants, as happens in thinning, but also requires the separation of a number of shoots that emerge from the multi-germ seed.

Skill was needed to separate the often-tangled clump of shoots, first of all by flicking them apart and then by cutting out the surplus ones. Failure to select and leave one shoot would ensure a harvest of intertwined spindly roots that would slip through any harvesting webs.

In England, the singling was often a two-phase operation, with workers going down the rows, 'gapping' the crop, leaving the more precise work to be carried out by a follower. Again, unlike the practice in Scotland, where a long-handled hoe was used, singling in England and throughout Europe was carried out with a short hoe that required the operator to stoop when working.

The tradition on Scottish arable farms was for the singling operation to be carried out by all the farm staff. Even the shepherd could not escape this duty as it was carried out during a quiet period in the sheep management year. While other parts of the sugar-beet growing world allocated breaks of rows to each employee, the Scottish tradition was to work in a line, each employee taking a row.

At the head of the line came the foreman, followed by

first horseman, or tractor man, then the second horseman, or tractor man, right down the line. After the main workers came the orramen, followed in turn by the bothy loons, halflins (not up to a full man's work) and students. Next to come were the women on the farm. Finally, the farm grieve completed the line-up.

The grieve's position was important as he could monitor the quality of the work, and where necessary, bark a reprimand to anyone doing less than a top-class job. The delivery of the reprimand was done loudly if a loon was the target, more discreetly if a tractor man or horseman was concerned, and only in private if the foreman himself was not up to scratch. Some farms operated a piecework system for singling outwith normal working hours. Piecework rates were generally agreed on each farm.

The skills of singling sugar beet were recognised and supported by the owners of the sugar-beet factory. Competitions were held, and from across Angus to Fife, men would come to compete in such competitions, which were often held in the evenings. Competition fields would be laid out so that each competitor would have to thin approximately 300 yards within a 2-hour period. Judges would inspect plant populations and freedom from weeds, and as the competitors took a well-earned bottle of beer, they would pick their champion.

Some of the skill in singling was removed in the 1960s with the arrival of selective weedkillers which could kill off some of the more difficult weeds, such as knotgrass and mayweed, which would tangle around the hoe blade. For Scottish growers, this advancement in husbandry came late in the day: less than half the 1967 crop was estimated to be sprayed with a selective weedkiller.

The biggest step forward in husbandry also came in the 1960s, with the arrival of monogerm seed. These seeds threw only one shoot; they could be sown to a stand, thus making the art and skill of the singler redundant.

Unlike turnips and beetroot that grow largely above the earth, sugar beet has long tap roots similar to most varieties of carrots. Harvesting is therefore more difficult and one of the first aids to lifting sugar beet came with the introduction of ploughs that eased the crop out of the ground.

In the early days, most of the hand harvesting was carried out by Irishmen lured by the cash on offer; the heavy, backbreaking work was left to these seasonal workers, to lift and then 'dad' together two beet to get rid of excess soil clinging to it. Most of this harvesting was done on piecework rates. For example, in 1937, Cupar NFU suggested farmers pay rates of sixpence, or 2.5 pence, per 100 yards for blanky (irregular) crops and, where the land was heavy, 'perhaps a little more'.

The beet were then laid neatly in rows; neatly so they could be easily and accurately topped with a long handled blade, or heuk. Next in the laborious process was hand loading onto carts. All of this work was carried out in the increasingly cold and muddy months of October, November and December: the campaign period.

As the black humour of the time went, the colder the day, the faster you worked. All because you did not want to hold onto beet that were glistening white with frost any longer than necessary. But again, mechanisation came along. First, there were the machines that removed the tops and then others which lifted the beet into rows.

Soon, this double operation was combined into one, with the added bonus of the beet being elevated directly into the trailers running alongside the harvesters. Back at the farm base, the trailers would be tipped up. In the early days, the beet then had to be hand graiped, either onto bigger trailers or lorries for delivery to the factory. Later, elevators were used. Often these would have a rotating drum or webs, so that some of the soil could be knocked off before the beet left the farm.

One problem mechanisation could not solve was the season of the year when harvesting took place. The sugar-beet crop could be taken too early. If this happened, the percentage of sugar in the beet was lower than it might have been. There was also the problem of mud on the public highways. Harvesting a root crop in November and December inevitably involves mud, and lots of it, and so the roads running between the fields and the farm steadings often resembled muddy tracks by nightfall. Motorists who may have spent time the previous weekend washing and polishing their cars did not appreciate this hazard on the carriageway. It was then that the orraman or the farm student became a roadman who spent his day scraping mud off the roads.

There were also problems at the sowing end of the sugar-beet season. If the seed went into the ground too early, a proportion of the crop would bolt, with the consequence of reduced sugar content, as described above. It was also a crop that required far more lime than any other on the traditional East of Scotland farm rotation. Farmers were wary of applying too much lime to their ground as the next potato crop might react to this by producing samples full of skin scab. However, the first husbandry problem faced by growers starting to grow sugar beet was the width of the rows. The practice on the Continent was for the rows to be only 14 in (34 cm) apart.

The horses used in the 1920s could not cope with these narrow lines. Other row-crops, such as potatoes and turnips, were planted or sown on much wider rows, between which the large feet of the horse could plod. Soon, the preferred row width was moved out to 24 in (60 cm).

The next issue was whether to sow the crop on the flat, or more traditionally on a ridge. The latter was preferred as it provided more scope for scraping away weeds as the crop grew. Against this was the problem that in a dry year there was less moisture in a ridge, thus preventing the young seedlings achieving maximum growth.

Initially the seeding was carried out with a horse-drawn seeder that covered two rows, but as soon as tractors came onto the scene, most mechanical seeders covered four rows. The first machines were just refinements of the turnip seeder, with seed driven out of a small hole by a revolving brush. This type of sowing inevitably left a veritable hedge of young plants in the row which would require singling.

Soon, the manufacturers came up with the idea of separating the seed. One type used holes on a belt that would only hold one seed. Another popular device had a wheel with indentations in the circumference. Again, each hole held a seed that would be propelled into the ground at the required distance from its neighbour. This helped the singling operation. After the crop emerged, scarifiers went down the rows, trimming the weeds from the sides of the ridge. The narrow row or band left also made singling easier.

For many involved in the fifty or so years of growing sugar beet in Scotland, life was a constant learning and improving process. One observer, who experienced the majority of those years, commented on the closure of the factory: 'We had a problem with multi-germ seed and now it is solved. We had a problem with weed control and now we have weedkillers. We used to have to do every job by hand and now it is mechanised. The only problem we have now is that we do not have a factory.'

Chapter 13

Forage Crops

AS the population moves further and further away from its rural roots, phrases born in the working country-side are drifting out of usage. One that is not heard a great deal nowadays is 'making hay while the sun shines'. For the countryman such a phrase is self-evident: if the weather is good, take advantage of it and secure the winter forage.

Great efforts were made during sunny spells in June to cut the grass. It was then allowed to dry before being stored away for feeding livestock over the hungry winter months.

Modern science can tell today's farmers just how good or how bad their forage is, simply by sampling and analys-ing it. The old system of sight and smell told our forebears just how much expensive, additional feeding they would have to purchase to augment the hay or silage.

Whenever there was a period of good weather forecast and the hay fields were ready to cut, it was all systems go. In the early years, hay was cut with a mower pulled by two horses. Ideally it was cut on a Friday and then, when par-tially dried, turned over on a Monday. After it was adjudged to be dry enough for storage, it would be rolled into rows. Next, it was gathered together with a tumbling tam, or hay sweep with long wooden tines, drawn behind a horse. The collected material was then dumped in a heap.

The second stage of drying and keeping came in

building small stacks, or coles. These had wooden or metal frames inside to prevent overheating of the still-drying hay. As the hay was thrown around the cole, the stack builder would level it out and trample it down. After the cole was topped out, it was raked down to ensure rain was shed by the stack. The 'sparty' ropes, made with esparto grass, were thrown over the top to prevent the wind opening the temporary stack to the elements. The ropes were secured at the base of the cole by twists of hay. Coles were allowed to weather and dry for 2–3 weeks before the hay was taken to the farm. This entailed hand forking the loose hay, although a number of farms had large wooden triangle lifters. After placing the triangle over the cole, metal forks were thrust under the hay and the whole lot lifted up by horse-drawn pulley. The load was then lowered onto a low flat cart, or paddock as they were called, backed under the lifter. At the stackyard a horse fork that could lift about 5 cwt, or 250 kg, helped build the large haystacks.

If the hay had not dried properly in the coles or if it had been a wet season, often some farmers added salt or ammonia to the stacks to help 'cure' the hay. Again, if the hay had not been properly dried, there was sometimes a risk of stacks heating up to such a degree they caught fire. A stackyard fire brought great excitement as workers tried to save other stacks in the vicinity. If the stacks heated at all, the hay lost a great deal of its goodness and any self-respecting beast would turn up its nose at it. Invariably, such hay had a 'fousty' smell rather than the sweet smell of well-made hay. Making good hay still requires a settled spell of good weather, and with no guarantee of that, there has been a recent shift to making silage as the main winter forage crop.

Interest in the possibility of 'ensiling grass', as it was termed, started in the 1930s. Lord Elgin from

Dunfermline put on a demonstration of the ensilage of grass in a concrete tower. The principle of making grass silage is based on fermentation and it is essential that the grass is properly consolidated to allow this process to take place. The early solution was the building of concrete towers and some of these, such as the one at Collairnie Farm, near Ladybank are still in existence. Workers had to go inside them and firmly trample down the silage as it came in on elevator chains. During the winter, as the cattle were fed, the silage had to be emptied by hand graip and thrown down a chute at the side of the silo.

Neither task was welcomed by the workers; during the filling, grass was pitched over the heads of the 'tramplers'. Then, at the emptying, those who climbed into the towers had to contend with the strong ammonia smell of well-made silage. At the filling stage, they may well have considered their counterparts in Canada and the US, where the farmers used to put an old horse in the tower when it was being filled. Gradually the horse rose in the tower as it compressed the silage. Once the silo was filled, the horse was shot, its work done. It was then tipped over the side.

Farmers who could not afford to build towers, or were tenants on farms where the landlords would not financially support such a venture, soon moved to making silage in a pit. Often, those pits were simply lined with old railway sleepers or set into an earth bank. The big problem with making silage is the potency of the liquor or wash that can seep from the ensiled grass. Sadly, the wildlife in many a burn and stream perished before this lesson was learned.

Nowadays, many farmers making silage do so by first baling the crop in the field and then wrapping the bales in polythene. The bales, which can weigh up to half a tonne, are then heaped together in the farm steadings, where occasionally they become a target for birds fascinated by their reflections in the black polythene. Initially there was resistance to feeding silage arising from a fear that it could sour the milk, and no doubt uncontrolled feeding may have tainted some samples. Today, however, almost every milk producer uses silage as the main winter forage.

In the early years, silage was cut like hay, with the wet or partially dried grass then carted to the silage tower or pit. But by the mid-1950s forage harvesters came along and they removed much of the hand labour apart from that of covering the pit with old tyres to keep it consolidated under the polythene sheet. Anyone who has had to throw tyres onto a clamp knows it to be a hard, hard job.

Turnips and other root crops

The arrival of turnips as a crop has been described as one of the pivotal factors in the Agricultural Revolution. The humble 'neep', as it is called in Scotland, allowed farmers to feed their stock through the winter. Before the English nobleman 'Turnip' Townshend (1674–1738) popularised the growing of the crop, livestock came through the winter in a half-starved condition. While his promotion of the turnip crop came from England, its importance was more significant in Scotland. Here, it became an important factor in the expansion of the Scottish beef and lamb industries. In Fife, large acreages of the crop were grown as farmers filled their turnip sheds for a winter of fattening cattle or over-wintered sheep on the fields.

In the early days, farmers needed a fair quantity of turnip seed as turnip-sowing barrows were fairly unrefined, compared to today's precision seeders that accurately drop a single seed. Rows of young turnip plants, sown by the early machines, would often resemble hedges. Before pesticides came along, these crops were frequently attacked by turnip flea beetles, which cheerfully chomped their way through succulent young leaves, killing the crop at an early stage. Seedsmen would then sell a replacement batch of seed to the farmer.

In pre-chemical days, farmers were so aggravated by the scourge of the flea beetle that they took drastic action. The flea's main characteristic was its ability to spring from plant to plant when disturbed. A cunning plan hatched by frustrated farmers saw a worker walk up and down the rows with a wheelbarrow. This carried a long pole, onto which was tied sacking that had been soaked in treacle. The theory was the insect would spring up in the air after being disturbed by the wheelbarrow and then find itself stuck to the treacly sacking.

The arrival of the first ever synthetic pesticide, DDT, in the 1950s gave farmers a powerful weapon against the flea, but few who used it realised the long-term damage this pesticide could cause. Nowadays, because consumers want to see fewer chemicals used, many turnip crops destined for consumption are covered with large porous polythene sheets that keep the flea at bay.

Today's turnip crops are sown 'to a stand', meaning they do not require any thinning of young plants, but in earlier years squads of workers would be seen at the thinning. In the East of Scotland they used long-handled hoes, but some farmers brought up in the West would straddle the row and then, in a kneeling position, move along it, thinning by hand.

The speed at which the crop could be thinned was often determined by weed levels, heavy weed infestation greatly slowing down the work. This was recognised as far back as 1922, when the Anstruther branch of the NFU decided the rate for thinning tonnage should be 2d per 100 yards, or approximately 1 pence per 100 metres. However, if the crop was especially difficult to thin, the Union considered the farmers could pay a little more.

Nowadays, the crop is all mechanically harvested, but in earlier times turnips were pulled by hand, with 4 rows hand-sheughed, or topped and tailed, into 1 harvested row. Two workers would take a couple of rows each and seemingly effortlessly pull, tail and top the turnip in one easy motion, swinging it into the middle of the rows. Carts would then be driven down between 2 of these rows. Sometimes the farmer would allow graips to be used for loading, but this was always considered a slower method than hand-filling. Loaded carts were then taken to the turnip shed or tipped into heaps in the field for feeding sheep.

Other forage crops

Apart from grass in its various forms, and root crops, a number of other forage crops have been grown in Fife over the past century. Crops of vetches and tares have now almost slipped out of sight, but 100 years ago, there were more than 1,000 acres grown in Fife, mostly for their protein content for cattle or pedigree sheep. Sometimes tares would be cut green and mixed with oats to form mashley, which was then fed to livestock.

One crop that seems to do well in many other parts of the world, but has not caught on in the UK is lucerne, or alfafa as it is called in the US. It grows well on lighter land, and over the years there have been a few enthusiasts. One Fife farmer who did decide to grow it found a market for the crop after Edinburgh Zoo wanted it to feed their elephants.

During the early years, field beans were often grown to feed working horses. In the days before sprays they were difficult to grow, with the bacterial black spot disease often decimating crops. Nowadays, growing beans comes into fashion whenever there is a surge in the price of protein feed, but this variable enthusiasm cannot remove another drawback to growing the crop: beans require a long growing season and it is not unusual to see combines going through crops in November and December. In a previous era, binders would be pulled out of their winter storage sheds to cut the crop.

One experiment to bring forward another forage option for Fife farmers that failed the test of time was carried out in 1928. Lord Cochrane of Cults tried to dry his oat and barley crops after cutting them green. He reported to the Fife Agricultural Association that the experiment had been successful, the corn was sold at a premium, the barley ground up for pig feed and the straw preferred by his cattle to 'good sweet hay', but the idea did not catch on, even though some cereal farmers might still consider such action following a particularly soggy harvest.

Chapter 14

Horsepower

THE stables were empty when I started to work on the farm as a schoolboy in the 1950s. Initially it was just odd jobs during the summer holidays, but I still sat proudly at the end of a line of workers.

As I waited for the farm grieve to bark or cough out the orders of the day to the men, I watched the rays of early-morning sunlight coming through the stable windows. They always seemed to be full of dancing dust.

The men, and they were mostly men, sat on wooden benches covered with an old jute sack that provided a slight softening to the cold wood. Perhaps, as they sat facing the empty stalls, many of them – former horsemen, themselves – would remember the days when the stable was full of the quiet clack of iron shoes on the cobbled floor and the harrumph of the horse clearing its throat as it waited to be harnessed up for the day.

Perhaps the 'now converted to tractor driving' men also remembered the early start to the horseman's day. Apart from a few months in mid-summer, this would start in darkness at around five o'clock in the morning as the horse was fed and watered before going out to work. The walk from the cottar, or bothy, would be in darkness unless there was a hurricane lamp to guide the way. Inside the stable, lights fuelled with paraffin were hung up at the end of the stalls.

The horses would then be fed and watered; the latter task required them to be harnessed and led over the close to the big, deep iron water troughs, or, as they were called, horse troughs. Having fed the horses, the men would then go back for their own breakfast. This would be similar to the horse's diet, as the first meal of the day was porridge made with oats.

Work in the summer started at seven o'clock but in winter, when there was less pressure, there would be a later start. The morning shift would last to midday, with a 15-minute break for 'piece time'. Today's tractor men often take a mere half-hour or even less at lunchtime when work presses, but the horse had to have a long break during the day to recover from the physical effort of the morning work. The midday break was often an hour and sometimes the argument was put that a 2-hour break was better. Normally, work finished at 5 p.m., unless the heat of harvest had taken hold on the farm. But that did not finish the horseman's day as even after 'lousin' time when the horse was safely stabled, fed and watered, the men would 'dander' or walk along the road to the stable in the evening just to check all was quiet and in order in the stable.

Day in, day out, that routine was carried on, with the exception of Sunday when one of the men took on all the feeding and watering duties.

The 'Sunday' men also had the responsibility of feeding the farmer's own horse – horses were not just used for work on the farm. Farmers required a light-legged horse to take them to meetings and to markets. This would normally pull a gig or small, open carriage. These horses were relied on, not only to know the road to the market but also to know the way home, if their masters had taken too much in the way of liquid refreshment.

Back in the stable, just as in later days, tractor men

would follow the same custom, the horsemen receiving their working orders for the day from the grieve. Again, in a practice adopted by tractor men, they left the stable according to their status on the farm.

If ploughing was on the cards, it was likely the field had first of all to be marked out in breaks, or riggs, by the foreman. He had the responsibility for ensuring these were correct, and to help in the task, his opening-out plough would also carry a long marker. This was a wooden pole, or 'lang marker', about 5 yards long, and as he and his horse went down one mark, it would gently mark out the next rig.

This sounds simple, but on the hills and side-lying fields in Fife there was a great art in ensuring the riggs were straight and true. Word would quickly spread through the neighbourhood if the ploughmen had to insert gushets, or inserts, into their work to get the riggs to match up.

Most of the ploughing was carried out with a pair of horse and this team could plough about 1 acre per day. That would provide time for the caring ploughman to lift gently any whaup's, or curlew's nest, from being turned over by the share of the plough. On heavy land sometimes a team of three horses was used, but if this happened, it required a special yoke between horse and plough. Horses always set their own pace, and indeed the pace of farming in those days was set by the speed at which they worked.

For the horseman the challenge in ploughing was not only to hold a straight line, but also to keep the correct depth with the plough. Some crops, such as potatoes, needed more depth of ploughing. It was always easier to lift the plough a little, but farmers were well aware of this.

In stony parts of North-East Fife, where there are many hidden rocky outcrops, the unsuspecting ploughman could be caught out by hitting a stone or rock that would send a judder through the plough shafts.

The skills involved and the importance of good ploughing were such that men would compete with each other to see who could plough the best rigg. This soon led to specialist ploughing competitions in the neighbourhood.

While experienced horsemen were entrusted with the ploughing and other skilled jobs, such as drilling and sowing, it was the work of the loon, or laddie, often with just a single yoke, to carry out the more mundane tasks. To him would fall the responsibility of harrowing crops, or taking seed and fertiliser to those carrying out the skilled jobs of sowing or spreading manure. At hay time, while the top men would cut and turn the hay, the loon would be trusted only with raking up wisps of hay that had been missed. In the early days, he would often use a tumbling tam, or rake that he would tip over once it had collected up a heap of hay or straw. At harvest, the first men would yoke their pair of horse into the binder, although in heavy conditions, they used 3 horses.

The next big task for the horses was the leading in of the grain crop. Carts would be converted to carry the loads of sheaves and the horse led by a youngster. Sometimes it just started and stopped between stooks of grain, with a shout from the horseman building his load, but this was a dangerous ploy as an unscheduled start or stop could pitch the builder off the cart.

In winter, the horses would be used to bring turnips in from the fields. Again, they would work between rows of pulled turnips, and start and stop according to the command of the team of workers busy throwing neeps onto the cart.

Another winter task for the farm horse was taking out the dung from the cattle courts. It would either be 'couped', or tipped, onto a midden, or if it was heading straight for the field, it was 'howked', or pulled, from the cart into heaps that would then be hand spread. If the going was hard, or if the cart taking bags of grain or potatoes to the train had hills to climb, it was not uncommon for a trace, or second horse, to be harnessed in front, giving two-horsepower rather than just one.

With all the direct action between man and beast, it was essential that commands were short and sharp. Almost every working horse had a single syllable name, such as Bob or Jock. Only pedigree animals carried long names indicating their breeding farm; one example being Baron of Buchlyvie, or more recently and locally to Fife, Collessie Cut Above.

As has been seen, in pre-tractor days almost all the power on the farms was supplied by horses, so it was essential to have the best ones possible. Sale catalogues listed the qualities of horses for sale and apart from the almost obligatory 'warranted in all farm work', there would be claims the horse was free from vice. This meant that it did not regularly kick or take a bite out of its handler.

Auctioneers were reputed to be reluctant to hold horse sales because of the types of people traditionally involved in such operations. The term 'horse trading' did not come into the language without its own pedigree, as all types of shenanigans took place when horses changed hands.

It was also imperative to keep farmhorses fit and the best ploughmen were those who tended their animals morning, noon and night. Regular small feeds were better than large ones. Some of the East Neuk farmers would take their workhorses down to the sea, as saltwater was supposed to be healthy for the horses. Throughout the country, this practice survives to the present day, with top racehorse owners taking their valuable animals down to train on the sands and into the breaking waves on the shore.

Even at the end of a working life, many of the work-horses provided cash for their owners as the firm, Grant of Broughty Ferry, used to buy a great number of horses for the European meat trade. These horses had completed their working life but still had one more service for man.

Nowadays, few would understand the life of a stallion man, but he played an important role in the countryside. Every spring, these experienced horsemen would take one of the top stallions around the farms in the district so that as many as possible of the working mares could be served by one of the top horses of the day.

There were a number of horse selection committees that nominated the preferred stallions. For instance, in 1933 the Fife Agricultural Society chose Woodbank Master from Allan Clark, Woodbank, for service in the Windygates area. The service fee was £1 10/-, or £1.50, with another £3 to be paid when the foal was born. In addition, a groom's fee of 2/6d, or 12.5 pence was paid. The Scottish Office was also keen to see the standard of horse raised and through their Board of Agriculture paid a subsidy of 25/-, or £1.25, per service. Jock McKenzie, who for years was the blacksmith at Ayton Smiddy, Newburgh, recalls his father being on the road every March, April and May, taking the chosen stallion around.

A route would be set out for the stallion to follow from farm to farm. Some farmers with superior females some-times took their mares to be served by noted sires. This was big business and the annual yearbook of the *Scottish Farmer* would carry pages and pages of advertisements for these prize-winning animals. One such trip describes the import-ance of this type of business. As a young man, Andrew Blair of Little Inch, Wormit was asked by his father to escort their prize-winning Clydesdale mare to Ayr on the West Coast of Scotland. There, she was to be served by

Dunure Footprint, one of the top names in the breed; the fee was a massive £60 per service, with a further £60 to be paid when the foal arrived. The journey involved walking the mare 6 miles from the farm at Little Inch to Kilmany Station; taking the train from there to Perth and then transferring to another train to Glasgow, followed by yet another train down into Ayrshire.

Thankfully, so popular was the stud at Dunure where Footprint did his work that it had its own siding. Arrival at the stud did not relieve young Andrew of his duties as the last words he had had from his father were, 'Don't take your eyes off her for a minute or they will use a poorer stallion.' Incidentally, it is recorded that Footprint served a mare every 2 hours for the 3-month mating season, which ran from March to May!

For most farmers in North-East Fife, Clydesdales were the breed of choice. They had a reputation for hard work and endurance. Pulling strength came from the hindquarters and deep, barrel-like chest, but most important for many of the old horsemen was the horse's contact with the ground: the old saying was, 'No feet, no horse'.

The worldwide importance of the Clydesdale in countries as far apart as Canada and Australia is illustrated by the export of more than 200 top mares and stallions in 1938 alone. Six of these came from Alan Clark at Woodbank, Windygates. Even to the present day, the Clydesdale Breed Society has more than 100 stallions registered for breeding, although their progeny are mainly bound for a life in the show ring rather than the ploughed furrow.

Some farmers preferred the Suffolk Punch with its red coat, however. They were introduced to the area in 1940, when there was a shortage of Clydesdales as they were reputed to have endurance and be hardy and strong.

Earlier in the 1930s there had been an importation of

Belgian Percherons, a breed still popular on the Continent, but the Anstruther branch of the NFU did not agree with this introduction of a foreign breed and proposed an import tariff should be imposed. No tariff was ever imposed nor was there a need to do so as the breed did not thrive in these northern climes.

The decline of the working horse

From a position prior to World War I, where almost everything that moved or turned on a farm was due to horsepower, there was a waning of importance. This was not a gradual decline, but rather one based on external circumstances. During the War, the Government could, and did requisition horses. As a result, much of Scotland's horsepower perished in war work, where these heavy draught horses took munitions and provisions to the front-line trenches.

An example of the shortages experienced during the latter years of that war come under a stuffy minute of the Cupar branch of the NFU. The branch deplored the fact that in England a special allocation of 250 tons of oats had been made for racehorses. This was, as one member remarked, a lot of oatmeal and it came at a time when the allowance of oats for workhorses was being cut as farmers strove to put as much onto the market as possible.

The resulting shortage of horses fit to carry out farmwork in the early post-war years gave considerable impetus to the budding tractor manufacturers. However, the Depression in the 1920s and 1930s slowed down the mechanical revolution as real horsepower was seen as being cheaper than that supplied by the internal combustion engine. By 1939, it was calculated only one-quarter of all horsepower on farms in the UK was of the mechanical variety.

World War II again put an impetus into mechanisation, this time not because workhorses were required for the war effort, but more so because their handlers, the horsemen, were conscripted and despatched to fight, leaving a shortage of able horsemen on the farms. After the War was over, another smaller battle took place between the horsemen and the increasing number of tractor drivers.

In 1945 Cupar branch of the NFU debated whether to pay horsemen a higher wage than tractor men, but a Mr Leslie of Tayport said he thought that farmers were paying for intelligence in respect of tractor drivers and they should not be paid less than the horsemen. Even if the horsemen came back to the land after the War, the progress of mechanisation was such that most farms had at least one tractor and this quickly multiplied as farmers and workers realised they did not need to feed and water their tractors seven days per week and at least twice every day.

Chapter 15

Machinery

IT should have been easy for I have driven tractors since my short trouser days. Admittedly the past two decades have slipped past without my spending any time behind a tractor steering wheel, but when I was asked to move one of the modern monsters around the stackyard, I thought, 'No problem.'

Getting into the tractor was actually much easier than it used to be, with neat little steps enabling entry to the comfort of the cab which is now an integral part of the tractor, unlike earlier models that were constructed as separate, even detachable parts. In complete contrast to the old days, the cab insulates the driver from the elements. In pre-cab days, tractor driving could be miserable when rain would fall steadily or a bitingly cold wind whipped right through old coats and headgear. The modern tractor seat was well upholstered, soft and springy, unlike the original metal effort covered with an old sack to help prevent a chilled rear and the ominous warning of piles.

So far, so good: there was a steering wheel but no gear stick in the middle of the floor above the main transmission shaft. All I could see were a number of smooth levers and handles, all within easy reach. I am no Luddite. Vaguely I recognised the global positioning technology and also controls for the linkage carrying the various implements. However, by the time I looked for the ignition key, I was also feeling that tractor-driving life had passed me by and the best solution would be to step down quietly and admit that I was not fully conversant with the latest technology.

I am sure my experience would be shared by anyone unfamiliar with the advances in agricultural machinery. It is hard to believe that 100 years ago, apart from the four-legged, hay-eating horsepower, the main power units on farms were stationary oil engines that drove mills, turnip cutters and grain bruisers. Some of these engines were giants of their day, with about 10-horsepower capacity, and they all had massive flywheels to maintain the necessary momentum for threshing or bruising grain. Many of the smaller two-stroke machines that 'phut-phutted' away, chopping up turnips or running hammer mills for barley or oats, demonstrated the early benefits of mechanisation. They were powered by oil gathered from the West Lothian shale deposits following the discovery by James 'Paraffin' Young (1811–83) that shale contained a valuable fuel.

Scotch oil, as it was originally called, was brought onto farms in 45-gallon drums in 1921. This fuel cost 11/5d per gallon, or just over 1 pence per litre, from the Scotch Oil Company, West Lothian. Another major use of this paraffin type oil was as fuel for the lamps used by the horsemen and dairymen charged with feeding or milking their animals before daylight.

In the early days, steam engines also came onto the farm. Contractors used them for ploughing or cultivation work.

As early as 1871, the Scottish Steam Cultivation Company had three rigs working in Fife and there were others, including an 8-horsepower machine working at Leslie. Incidentally, the use of horsepower as a standard measurement or power did not occur until the 1920s, following complaints of spurious claims by dealers about the capabilities of their machinery.

In 1914, A. & R. Brown of Colinsburgh came to Drumrack, Anstruther to cultivate ground. In total, they covered some 14 acres, for which they charged 13/-, or £0.65, per acre. Other contractors worked steam-ploughing rigs: one such machine was operated by the Murray family from Guardbridge on ground at Easter Kincaple, where a steam engine was set at each end of the field to be ploughed. The plough would be dragged back and forth on long wire pulleys between the two machines. Because the furrows were turned in a certain direction, the plough itself was an early version of a reversible plough, with three or four furrows in the ground and an identical number held in the air on one trip and then dropped into the ground to work the return journey. The steam engines owned by the Murray family were once used to help pull a Hawker Hunter aircraft from the Eden estuary. Either the pilot had misjudged the location of the nearby Leuchars airfield, or had ditched into the muddy shoreline.

Steam engines were also the main power unit for the mobile threshing units that travelled the countryside. Fife had a number of these contractors operating in the winter months round the rural areas. However, the days of the steam engine were cut short because of inflexibility and high running costs.

The introduction of the motorcar brought with it ambitious inventors who wanted to convert these machines with their internal combustion engines into workhorses for the agricultural industry. Many of the early tractors were based on the same principles as cars. The first major car manufacturer involved in tractor production was Henry Ford, the man who in 1908 pioneered the production line system of manufacture in the USA. Other car manufacturers tried to follow suit in bringing out tractors. Notable among them was Austin, who started tractor production in 1917.

Prior to the car companies becoming involved, the early days of tractor manufacture saw a number of entrepreneurs and inventors take up tractor development as a challenge and an opportunity. Typical of this first generation of tractor makers was Professor John Scott, of Duddingston near Edinburgh. He was a visionary who believed tractors should be able to carry implements in front of them rather than just trailing behind, as all his early competitors believed. Professor Scott also produced an early version of a linkage that could lift rear-mounted equipment out of the ground. Forty years later, this idea helped make the fortune of Harry Ferguson with his ubiquitous wee grey 'Fergies'. Sadly, in the early 1900s when he toured the country promoting his invention, Scott ran into great opposition from a farming establishment that could not see beyond the merits of horses.

Fortunately, other tractor manufacturers continued their efforts, and in the later years of World War I, they made a breakthrough. By this time, many farms were short of draught horses; they had been requisitioned by the Army and tractors had to fill the power gap back on the farm. A large Ministry of Munitions importation of Ford tractors from the production lines in the US gave an indication of future manufacturing methods. The introduction of these tractors was not the first example of the military providing a pathway to the future for farming. Throughout the War, they supported the baling of hay by steam-power because the traditional loose bunches of hay could not be transported to war zones to feed the horses.

Some idea of the scale of interest in tractors after World War I came with the Great Exhibition of Motor Traction in Lincolnshire in 1920. Some 55 tractors were put through their paces on the flat Lincolnshire land. Included in this impressive cast list was the Glasgow, a revolutionary three-wheeled tractor, where all three wheels were powered. Despite the Glasgow receiving rave reviews, production by the D.L. Company in Motherwell never took off, possibly because the farming industry slumped into recession in the 1920s and 1930s and farmers turned their backs on mechanical power. In those years of the Depression, horses came back into favour as they ate home produced hay and did not require expensive fuel. That is not to say that tractors disappeared from farms; it was just that they did not replace horses as quickly as is sometimes thought.

The first-ever tractor at Drumrack farm, Anstruther was an International Harvester Titan, which in 1917 cost £170. It was bought after performing at a ploughing demonstration when it showed that it only required 1.5 gallons, or 5 litres, of paraffin per ploughed acre. This paraffin or petrol/paraffin mix was standard in early tractors – diesel technology did not advance to a level where it could be commercially used until the second half of the twentieth century. Incidentally, the impressive ploughing carried out by the Titan was similar to all other tractor ploughing in the early days, a two-man job. The 'sitting' ploughman, perched on the implement, was required to control the levers that kept the plough straight and true – there was only a pulling chain between tractor and implement.

All ploughing carried out today sees a direct fixed linkage between tractor and plough, with the tractor driver totally responsible for the quality of the finished work.

Early machinery selling arrangements were far more flexible than the current system that sees dealers tied tightly into tractor and machinery manufacturing companies. Now, they are normally only allowed to promote and sell models of that marque. In the old days, local companies could source a specific make of tractor, even if they were not the recognised dealers. A network of small companies in Fife sold tractors and other farm machinery. Among them were Gillies & Henderson, J.B.W. Smith and A. & J. Bowen. Others, such as the Caledonian company, had a bigger franchise area in selling Caterpillar tracked vehicles.

The big advance for tractors came with the need for more production from the land and the war years saw a doubling in the number in Scotland. By 1942, it was reported some large farms were 'entirely wrought by machinery without the use of a single horse.' Along with the conversion into tractor power, tractor and implement dealers in agriculture flourished in the years from the end of World War II right up until the end of the 1970s. Many of these were not only agents for a make of tractor, but also had a range of equipment suppliers.

During the last two decades of the twentieth century, the agricultural machinery market has become a global trading place. International companies now dominate the market and this move towards globalisation has left many of the smaller, local dealers without franchises. There are still around ten main tractor suppliers in Fife, but all are selling vehicles made in other parts of the world.

The range of machinery on the farm
To grasp an idea on the scale of change in farm machinery in the past 100 years, it is only necessary to compare a farm sale catalogue from the beginning of the century with a similar, modern-day sale.

The roup roll (the brief description of all the articles at a sale along with buyers' names and prices paid) for Lower

Luthrie and Carphin farms in 1897 showed the equipment needed to run a 600-acre mixed arable and livestock farm. Four reaping machines were the forerunners of the binder. These had no method of tying or binding the cut grain, but they did remove the onerous task of hand scything the crops. Nowadays it is possible that a farm of this size might have a combine, but it is much more likely harvest work will be contracted out to a neighbour whose machine has spare capacity. Some of today's bigger combines can easily harvest 100 acres per day, but then these modern harvesters can cost well in excess of £250,000.

There were no less than seven single-furrow horse-ploughs lined up for sale at Luthrie. In the early days, local blacksmiths could make their names by hammering out a plough that suited the land about. However, it was not long before the plough-making reputation of some black-smiths spread beyond their own parishes. For example, ploughs from Cameron of Tullymet in Perthshire or Barrowman from Saline were spread around the country-side. There was a big breakthrough in plough technology in the 1920s when Oliver, a manufacturer, introduced a plough with cast-iron replaceable parts.

Other manufacturers followed with their own version of replacing the wearing parts on ploughs. This advance was welcomed by farmers who had previously to take their ploughs to the local blacksmith when shares or boards were worn; now, they just purchased replacement parts. Soon major manufacturers were in on the act; prominent among them was Ransomes from Ipswich who gained a good name, and many a ploughing match victory was taken by one of these ploughs.

Today, a farm sale will include one or possibly two large multi-furrow reversible ploughs. These work from one side of the field to the other without the old system of drawing feerings or breaks for the horse ploughman. Today's plough will also be a 4-, 5- or 6-furrow machine capable of ploughing more in minutes than an old plough-man could in a day.

More old-time cultivation equipment followed, with drill ploughs for turnips and opening up potato drills, grub-bers for loosening the soil and helping to prepare a seedbed and two rollers to consolidate the lighter land and break down clods on the heavier soil. One was an iron roller and was meant for the heavy work, and there was also a wooden roller used in less demanding conditions.

The roup roll continued with a range of iron harrows. Various local blacksmiths filled any spare time between shoeing horses by making harrows and they all used their own designs. Thus, the sale records a 'range of harrows'.

Modern-day cultivation equipment combines a number of processes in one working unit. On a cereal-growing farm, these are also combined with a grain seeder to make a 'one-pass' machine. As the name suggests, this cultivates and sows at the same time. The scale and speed of the com-bined cultivator and seeder makes nonsense of the seed fiddle that appeared all those years ago at the roup.

The auctioneer of a century ago moved onto the horse sweep. This was called a 'tumbling tam' and it was used to gather hay together into heaps, ready for the men to fork into small stacks, or coles. There was also a tumbling rake that had a valuable role in the harvest field of gathering up the last wisps of straw. Why did they gain the nickname 'tumbling tam'? The answer is simple: after collecting the hay or straw, the operator would tip, or tumble, the machine over, leaving its load behind. A mechanical mower used to cut the hay was also listed. Today's modern forage-making equipment often combines a number of functions. Mowers may cut, but they also crimp; forage harvesters not only pick up silage, they also blow it into a bulk trailer.

The old-time congregation, with the auctioneer taking the central role, then moved onto the carts. Most popular was the coup cart, with its short body made of wood. It was made for pulling by one horse. The shafts would come along either side and the body was built to hold approximately 1 ton of produce. This was reckoned to be the appropriate weight for a draught horse to pull.

The coup cart was clever insofar that the losing of a pin and a hefty heave at the front of the cart would tip the load of potatoes, turnips or dung. Although I did not know it at the time, seeing the couping, or tipping, provided me with one of my first lessons in physics as the pivotal point was just off-centre. Local joiners would link up with blacksmiths to make these carts. At Kames Smiddy, south of Cupar, was an example of these complementary businesses working together. In addition to the carts themselves, there were also special sides called 'flakes', which, after the removal of the normal wooden side, would be lifted over the body of the cart. Then there was a hay bogie, or low, flat cart used specially for bringing home heaps of newly made hay.

The difference between the cartage on display a century ago and today is largely down to scale. A modern farm sale might involve a large bulk trailer capable of carrying around 10 tonnes of grain or potatoes. Today's sale items probably include flat body trailers, recycled from the haulage industry to carry a dozen or so 1-tonne potato crates or twenty or so large bales of straw.

Again, shifting back in time, a range of horse harness was listed. The roup roll shows a dozen swingletrees used to catch the lead reins and transfer the two pulling lines into the chain that was attached to the machine in use. Halters, heims, brechams, barn collars and saddles were all lumped together in a lot marked down under the heading 'harness'. There must have been a fair amount of leather and chain in that heap because the purchaser paid £5 for it.

There is no modern equivalent of the harness unless one considers the spare tyres, oil drums and extra counter-balancing weights that go in front of the tractors.

While the old roup would move to the stables to sell the horse, modern horsepower-tractors are lined up in prime position in the machinery lines, starting with the most powerful and expensive. However, today's tractor market can throw up unexpected bids. The latest model oozing with electronic gadgetry may have been the main power unit on the farm, but it can often be sticky to sell as second-hand buyers are somewhat wary of equipment that cannot easily be repaired and so a more mechanical, albeit older model can sometimes see competitive bidding as it has greater universality and dependability.

Hand tools

All farm sales start at the barn door. Hand tools and machinery spares are passed out to the auctioneer, who stands on an old cart or upturned potato box to see the bids. There is always a range of hand tools, such as hoes or clats, as they used to be called. Essential in pre-weed spraying days, they would be used to hand-hoe potatoes and again at turnip thinning time. All the hand tools would have shiny shafts and handles, worn smooth through years of use.

Decades later, when hoes were still necessary on the farm, I recall that at beet singling time every man had his own hoe. Part of this was down to the belief that they were all minutely different and even the slightest bend or twist in the shaft would end up with slightly less than perfection in the finished article. More prosaically, I recall not wanting to use the farm grieve's hoe as he smoked constantly and equally regularly spat on his hands. The result was an unpleasant, dark-stained hoe shaft.

In fact, on most farms in the old days, each man had his own set of tools; his spade, shovel and fork were hung up on his hooks in the stable where they started work each day. Spades and shovels still appear from the barn door at farm sales, but they are used far less now than in the days when every action required physical effort. Then there would be other hand tools, such as dung hauks, which were used to empty the coup carts, or the special ditching spades that reduced the amount of work needed in laying new drains.

In the old days, if draining was needed on a farm then the only piece of equipment to hand was the spade, albeit a special narrow draining spade that saved effort in digging a trench. A working day for a man at draining was reckoned to be 1 chain, or 20 metres, at a depth of 3 feet or almost 1 metre. For today's farm sale, draining equipment will be seen out in the field. A hydraulically driven shovel attached to the back end of a tractor can quickly dig out the necessary track and carry out any other earth-moving required around the farm.

There were other hand tools, too. These included items such as tapners, or knives, for taking the tops and tails of the all-important turnip crop and long-handled hedge bills for trimming hedges and cutting down thistles.

Next to be handed out of the barn door on that sale day 100 years ago were sets of potato riddles. These not only sized the crop before it was sold, but with a flick of the wrist, the potatoes on the riddle could be turned over and any rotten ones picked off. The next stage in the evolution of sorting out potatoes was the grader driven by a small petrol engine. It had shaking or revolving riddles, and inspection tables fitted with wooden revolving rollers and hooks for hanging sacks on.

It would be exceptional today for a farm roup to have a potato grader as most of the crop goes off to centralised stores, where massive machines with electronic gadgetry sort the crop out more efficiently than would ever be possible with the human brain or eye. Instead of having grading machinery, sales at farms where potatoes are grown will see stacks of 1-tonne wooden crates used for transport and storage.

Scythes now occupy the dusty corners of old tool sheds, and even if anyone can get them down from the rafters where they were always stored, it is fairly unlikely any handler would be able to replicate the swinging motion required to work them properly. Even less likely would be the ability of modern man to sharpen such a fierce piece of machinery without the accidental removal of a digit or two. The old method was a quick spit on the sharpening stone followed by a swift and rhythmic motion of back and forward along the blade. At the same time, the stone would be moved seamlessly from one side to the other.

On that day more than a century ago, scythes and sharpening stones had the same value as the potato riddles. Today they only have rarity value, often bought to hang on walls as part of a nod to a bygone age, or even as a decoration for a village pub.

On livestock farms, fences were not so good as they are nowadays and so, when fence stobs (posts) and fencing were under the hammer, they were cleared quickly and at good prices. The fencing could not be completed without the purchase of a mell for hammering in the pointed posts. This lump of iron could weigh up to 14 lb or 5.5 kg and was often used by the ploughmen as a cheap substitute for throwing the hammer, as seen at the Highland Games. Also coming out of the barn door with the rest of the fencing gear would be the guddle, a much-underrated implement. This is a heavy metal spike that greatly reduces the use of the mell by 'guddling' out a hole for the stobs.

The sale was still not complete as there were important items such as stack props or wooden poles that could be used to prevent an embarrassing 'Tower of Pisa' tilt on the straw stacks.

The stacks were built on staithels, or mushroom shaped stones that prevented rats and mice from climbing into them. Six or eight staithels formed the perimeter of a circle and then a metal or wooden frame went on top, keeping the straw a couple of feet from the ground. The stones provided an effective barrier to vermin and the raised bed ensured no loss of grain occurred through dampness.

At the Carphin roup the staithels were knocked down for a few pieces of silver, and even in a 1930 roup roll, they only made 10/-, or 50 pence. Staithels are still put up for sale at farm roups, but buyers can expect to pay more than £100 for each one. They have become very popular as garden features although their owners will never have to deal with a hungry rat trying to climb them.

Few nowadays would recognise a thrawcrook, but these hand-operated tools for making ropes were very important to our predecessors. Making ropes to help hold down the thatch on the straw stacks was a skill honed on wet days in the farm loft. Gentle turning of the hooks ensured the proper twisting necessary for a good rope. Any other speed or even the loon letting go of the rope at a vital time would ensure a word or two of correction from the farm grieve.

A fairly new arrival on the farm machinery list a century ago was the steelyard, which was used for weighing grain and potatoes. Prior to its arrival, grain was despatched by volume with bushels the main unit of measurement. Still on items dealing with the grain crop, there was even a sack lifter with chains and ratchets that helped lift the 110-kilo sacks used in those backbreaking, knee-bending days.

To the untrained eye, there is little to see in the feering poles that came out of the barn door. They were a bundle of poles about 2 metres tall, with a sharp end to help fix them into the ground. Feering poles were essential as the ground to be ploughed not only needed to be measured out, but the drawing of the first furrow had to be straight and true. Before the furrow was drawn, these poles were lined up in the field. Usually painted in some unusual colour such as red, they could be seen against any country background. The setting-out of the feering poles was often fraught with problems: a flat field was relatively easy, but in one with a ridge running along the middle the ploughman could not see the other end and so he relied heavily on his feering poles. Setting them out in such a situation required a knowledge of how to keep three poles within sightline at all times.

Still more items came out from the depths of the shed. These included oil pouries, or cans, and grease guns. Farm machinery may have been fairly basic, but it needed regular maintenance with a squirt or two of oil, or a 'slagger' of grease to keep the bearings running smoothly.

With a fair bit of effort, wooden ladders were produced from the barn door and then after careful inspection, to ensure most of the rungs were in place and not too many of them tied up with wire, the bids were taken.

Domestic items were also included in the sale. Beds from the bothy may have seemed to be a fixed item that belonged to the landlord, but they too were up for sale. Also on offer were the clothes poles from the cottages. Perhaps the sale of this low-value item had more to do with the relationship between tenant and landlord than with raising every penny possible. In the days when no electricity was available, there was a big demand for barn lanterns fuelled by paraffin. They lit the byres for the early-morning milking, or the stables as the horses were given their first meal of the day.

The crowd at the sale then moved round to the loft, where the first item under the hammer was a fanner used to winnow, or separate, the chaff from the grain. In the early years of the last century, this was still the best method of ensuring a sample of good grain, free from light ears and weed seeds.

Next up was a cake breaker. Up until the 1960s, slabs of protein-rich cake from exotic countries would come into the UK. Before they could be fed to any of the cattle, they had to be broken down into edible chunks. The mechanism was simple: two rollers complete with 'slab-breaking' knobs and bumps turned by a large handle.

The arrival of a lorry-load of slabs of this cake on Craigie Farm, Leuchars during a particularly wet spell in the early half of the last century saw an innovative, though very wasteful solution to the puddles in the farmyard, with the men using the slabs of cake as steppingstones.

The sale was finished with a 'scutter' of articles required for keeping hens. Items such as coups, feeders and zinc water buckets were bundled together and sold. They may have been just bits and bobs, but the mixed farm in those days always had a few hens around.

Power and communication down on the farm

As today's farmer switches on lights and presses the starter button on machinery, he or she might also find the ever-present mobile phone sounding its alert from deep inside a pocket.

Sixteen year old David Russell is as proud as Punch as he stands alongside the foreman's pair of horses at Lower Luthrie Farm, Cupar in 1922. *(Andrew Arbuckle)*

Seagulls flock onto the newly turned earth following Tom Lorimer as he ploughed at Logie farm, Newburgh. There were tractors on farms in 1935 but the single-furrow plough pulled by two horses was still the most common method of cultivation. *(Tom Pearson)*

Ploughing was competitive and on ploughing match days fields were marked out in plots so that the quality of ploughing could be judged. Awards were also given for the best turned out horse and sometimes even for the best looking ploughman. *(Tom Pearson)*

Sowing turnips in the 1930s, with a horse-drawn plough to make the ridges and a horse-drawn seeder to drop the little seeds into the ridge. *(Tom Pearson)*

A three-horse binder working in the 1930s. Even although the crops were much lighter then, it often required three horsepower to turn the various wheels of the binder. *(Tom Pearson)*

Tractor power takes over but the binder still requires a man on the machine to work the many levers. *(Tom Pearson)*

Stooking sheaves at Berryhill Farm, Newburgh in the late 1940s. The sheaves of grain made by the binder were left to ripen and dry in stooks made with ten or twelve sheaves. *(Tom Pearson)*

Bringing the harvest home at Drumrack Farm, Anstruther in the early 1930s. The sheaves were hand forked onto the trailers before being taken to the stackyard. *(Henry Watson)*

Building the straw stacks in the Drumrack stackyard with the sheaves being forked to the stack builder. *(Henry Watson)*

A fine row of stacks of Nether Strathkinnes, St Andrews. They are thatched, ready for winter, and neatly trimmed. Close to a busy road, the craftsmanship would have been admired by many neighbours. *(Tom Pearson)*

The labour intensive work of threshing grain. The funnel of the steam engine driving the mill is seen beyond the mill itself. In the foreground, the threshed straw is taken away in bunches for bedding livestock or for protecting clamps of potatoes from frost.
(DC Thomson)

He may be smiling but James Mackie's legs are buckling as he lifts a sack of grain weighing over 100 kilos at a threshing on Back Dykes farm, Crail, in the late 1950s.
(Tom Pearson)

Not combining grain in the Prairies but four Massey 500 combines belonging to contractors, A & J Black, working in a barley field on Logie Farm in the 1960s. *(Andrew Arbuckle)*

At Drumrack Farm the grain harvested by the Claas combine is transferred in bulk to a trailer pulled by an International tractor running alongside. It is the early 1980s and gender equality was established with both vehicles being driven by women. *(Henry Watson)*

Potato planting required a great deal of horsepower as this photograph from Rankeillour demonstrates. Drills were drawn, then split and fertiliser applied before planting. *(Tom Pearson)*

In the 1950s, planting potatoes by hand from 'brats' made from jute sacks. As the women work their way down the drills, the men bring more supplies of potatoes forward. *(DC Thomson)*

An early mechanical potato planter working at Parkhill Farm, Newburgh in the 1960s. To keep the correct spacing in the ground, every time a bell rang, Sandy Pearson and Lou Aitken had to drop a potato into the spout. *(Tom Pearson)*

Hand-lifting potatoes at Scotscraig Farm, Tayport in the early 1980s. A big squad of pickers are kept going with two two-row elevator diggers while the potatoes are being filled into one-tonne boxes to be taken away and stored. *(DC Thomson)*

Machine harvesting potatoes at Fliskmillan Farm, Newburgh in the early 1980s with a Johnston Faun single-row harvester. Hand pickers worked under the canopy removing the clods and stones before the crop was transferred to a trailer running alongside. *(DC Thomson)*

Pulling flax at Balmeadowside, Newburgh in the 1930s. This work was sore on the hands as the plants were deep rooted. The hats or 'mutches' were commonly worn by women in those days. *(Tom Pearson)*

Modern Mather & Platt pea viners working in front of the Lomond Hills in Fife in the 1990s. These combine-harvest the crop before transferring it in bulk to lorries. *(East of Scotland Growers)*

A trailer load of harvested peas is transferred onto a lorry before going off to the freezing factory. The harvesting operation was programmed around a number of farms. On this occasion, the action is taking place at Ballinbreich, Newburgh, in the late 1980s. *(Andrew Arbuckle)*

A dirty job on a windy day but the application of lime in the first half of last century was a major factor in improving crop outputs. *(DC Thomson)*

Planting Brussels sprouts at Reedieleys Farm, Auchtermuchty in the 1990s, with hand planters feeding specially grown seedlings into the machine. Their work is being checked by the two workers on foot whose job it was to fill any gaps. *(DC Thomson)*

Preparing sugar beet for singling in the 1950s. The steerage hoe trims away the soil on either side leaving a narrow ridge where the plants have to be thinned; a skilful job where error and removal of plants would incur the wrath of the farmer and foreman. *(Peter Small)*

Hand thinning sugar beet at Back Dykes, Crail, again in the 1950s. Taking the lead is the foreman with the rest of the farm staff each taking a row behind. Because thinning took all the labour force on the farm, under school age children would often accompany their mother in the field. *(Tom Pearson)*

On steep land, a trace, or second tractor was brought in to help. In this case, a Massey tractor helps an early attempt at the mechanical harvesting of sugar beet with a Catchpole machine at Fliskmillan Farm, Newburgh, in 1950. *(Peter Small)*

Working land in 1934 with a Fordson N tractor pulling a Wilder pitch pole harrow at South Lambieletham farm, St Andrews. The tines of the harrow could be reversed to help gather couch. *(John MacNiven)*

Loading turnips into a coup cart in the 1950s. This was back-breaking work as the turnips had to be hand loaded to prevent damage. *(Tom Pearson)*

A prize-winning crop of turnips at Hilton of Carslogie, Cupar, in the late 1940s. The production of big yields of Swedes was important for farmers who fed them to cattle. *(Pat Laird)*

A turnip pit at Daftmill, Cupar. After the coup carts had deposited their load, the pit was strawed as protection against frost. Care had to be taken in handling the crop. *(Tom Pearson)*

Piece time at potato lifting in the early 1980s. A fifteen-minute break was normally taken about 9.30 a.m. with another ten minute break at 3 p.m. *(DC Thomson)*

A busy time at the berries with a queue of full buckets waiting to be weighed in before being transferred to the barrels on the trailer. Payment was made after the berries were weighed. The little girl is checking to see she gets the right amount of cash. *(Andrew Arbuckle)*

In 1934, Jack MacNiven, South Lambieletham, St Andrews, and Jimmy Carstairs get the manure barrow ready for applying fertilisers. An unnamed boy is left to hold the horse. *(John MacNiven)*

The traditional method of fattening lambs 'on the break' with the sheep being fenced off in a section of the field so that they clean up all the turnips before moving to a fresh break. *(Tom Pearson)*

An empty stackyard at Parkhill, Newburgh. The staithels kept the harvested sheaves off the ground. Notice the upside down cups to prevent vermin climbing up into the stack. *(Tom Pearson)*

Bringing home the hay in 1936 with an International Farmall tractor pulling a trailer full of loose hay into Drumrack farm steading. *(Henry Watson)*

Loose hay is transferred to a larger stack with the help of a hay fork. Using a horse-drawn pulley common in the 1920s, this fork would lift a large heap of hay onto the stack. The builder would then level and trample it down.
(Henry Watson)

Hand milking cows in a modern byre at Balbirnie Farm, Markinch, in the early 1950s. A good milker could hand milk eight cows in each of the twice a day. *(DC Thomson)*

The delivery of fresh milk to the cities involved rail transport. In this case, milk churns are being handled in Dundee station in the early 1920s after arriving on a train from Fife. *(DC Thomson)*

A daily duty for the farmer's wife was feeding the hens with the scraps of food from the farmhouse or using some corn or barley grown on the farm. The hens provided a regular supply of eggs – and meat if the hen stopped laying.
(Tom Pearson)

A typical scene of fattening cattle on Brigton farm, St Andrews, in the East Neuk of Fife. This photograph was taken in 1934 when horned cattle were still commonplace. *(John MacNiven)*

A busy sale day at Cupar mart in the 1950s. Jimmy Garland in the auctioneer's box with a pen of Suffolk cross lambs from Peter Skimming, Airdrie Farm, Anstruther, being sold. *(Garland)*

Typical of the constant move from farm to farm in the first part of last century. In this case Andrew Williamson, Bank farm, Newburgh is carrying out the flitting of a farm worker's worldly possessions in his cart as it passes down through Newburgh High Street. *(Tom Pearson)*

A farm sale or roup at Newton of Nydie, Strathkinnes, in the 1950s with the auctioneer the centre of attention of all the bonneted farmers standing around. *(DC Thomson)*

Preparing for a farm sale at East Flisk, Newburgh in 1999 with the machinery all laid out in rows. As this farm grew raspberries it had a 'berry bus' to bring the pickers to the farm. *(Tom Pearson)*

Going off to Fife show in 1945. The farm staff at Fernie Mill, Cupar may not have realised it at the time but their trailer had a slow puncture and they had to stop for repairs before they got to the show. *(Tom Pearson)*

A rare photograph of the Highland Show in Cupar in 1912. The judging of the all-important horse championship is underway in the main ring. *(Tom Pearson)*

All this instant technology would astonish people from an era where considerable effort went into providing any artificial light or non-human power. They would be totally mystified by the thought of communicating using a small hand-held phone with those you could not see. And yet as recently as 1932, Mr E. M. Hood from the British Electrical Development Association addressed members of the Fife Agricultural Society on the benefits of electricity.

The invitation, wonderfully predicting the possibilities of instant power and light, stated, 'electricity is expected to play an important part in assisting farmers in the future.' True to the advance billing, Mr Hood pointed to benefits such as a reduction in milk spillage in the dark, early-morning milkings. In addition, he pointed out to his audience there was far less chance of fires being caused by upset paraffin lights.

Another promised bonus was an increase in egg-laying through 'artificial light': it had just been proven that hens which traditionally laid their eggs in the summer months could be fooled into thinking they should keep laying as long as there was light. Mr Hood stated that the biggest poultry farm in the world, the Buttercup Dairy company on the outskirts of Edinburgh, was using electricity in the units for its 150,000 laying hens.

From that meeting, interest in electricity on the farm grew. In the 1930s it was installed in many farms by the local Fife Electricity Board, while those closer to towns or lying between them were more fortunate as they could often be linked to the power lines crossing their land. In more outlying properties, farmers bought their own generators. Farms and cottages were served by these diesel engines that steadily worked away, producing electricity, albeit with a lower voltage than 'mains'.

In the 1950s I recall my father going out each night to check the generator. Any time it hiccupped, candles and tilley lamps were pulled out of the cupboard.

Some farms decided gas provided a better option and Rumgally, Cupar bought a system that produced gas for light. The contraption consisted of two large water tanks, into which calcium carbide was added; the resulting chemical reaction created a gas that was then used for light, similar to early versions of bicycle lights.

Prior to electricity coming onto farms, coal was the main power source as it fuelled the steam engines that turned the early threshing mills. Fife had few water-powered mills: many of the arable, cereal-growing farms are on relatively flat land and watercourses around these areas could not drive the threshing mills.

While most of the big coalmines in Fife were in the west of the county, a number of smaller ones were in the east. A drift mine at Radernie, in the centre of North-East Fife, closed in the late 1940s. Prior to that, there were coal workings at a number of locations.

In the coal shortage after World War II, the National Coal Board proposed an investigation into up to eight new sites in the East Neuk of Fife. The local NFU objected to these proposals. Ironically, only a year or two before, the same branch of the Union complained about the shortage of coal. It considered the matter so serious that, in some cases, horses could not be shod because the blacksmith did not have enough fuel. No development of these potential coalfields ever took place but that was more likely due to the small, possibly uneconomic, scale of the sites rather than the Union's riposte.

The mobile phone has already been credited with the demise of door-to-door farm salesmen. A great deal of today's business is conducted from the tractor seat while the ploughing, sowing and harvesting go on. Yet, less than eight decades ago, many farms, especially in remote areas, decided they could not afford to be connected to the newly arrived telephone system. In the mid-1930s, it was still possible for one commentator to report, 'telephone facilities are not found widely.' Two decades later, in 1950, there were twenty holdings in the Pitcorthie area, near Anstruther, with not one telephone among them, so they asked for a kiosk to be provided.

Other areas closer to main towns were quicker to discuss the potential for the telephone. In 1921, J. L. Anderson of Dairsie considered the telephone would be a great boon to farmers, especially when they had to deal with the local station in ordering grain sacks or potato wagons. His enthusiasm was misplaced, as it was pointed out that many of the stations did not have telephones.

The cost of such a service was also a deterrent to many, and a single subscriber in the Dairsie area would face an annual bill of £35. However, if they could get up to twelve subscribers on a party line that cost would drop by up to 50%. Another option available in the days when phone connections had to be carried out manually was an all-night service that would cost an additional £6 per annum.

This is all a far cry from today's immediate and easy communication system, although it is not without its own problems. During a spell baling straw, one East Neuk farmer lost his mobile phone when it slipped from the tractor seat into the path of the baler. There was only one solution: he went back to the farmhouse and asked his wife to phone the mobile. Shortly afterwards, his neighbours could see him going around the field, listening to every bale.

Chapter 16

Beef Cattle

PART of the Sunday-afternoon ritual in my 1950s youth was the trip around the countryside. Although then a concept I was unaware of, this was the farming equivalent of low-level industrial spying. Not that my father was alone in this activity. Looking over the farm dykes, through the fences and over the hedges, was a popular and often educational pastime. At the beginning of the century, tenant farmers and landlords would frequently use the height of their horse gigs, or the view from their saddles to observe the goings-on at all the farms in the neighbourhood. Workers would also be seen on their bicycles, checking up on the quality of ploughing, or the bigging of the stacks of their colleagues on other farms so that boasting rights could be established at the next social gathering.

Even in the present day, the observation of work on another farm is the mark of all who live in the country. This practice is called the 'rotating bunnet' or 'rural lighthouse' syndrome, as anyone in the vehicle following can see the heads turning this way and that on all the rural roads. So, bundled into the family car, siblings squabbling as to who had the side seat and who was to sit on the humpy bit in the middle, we progressed at a fairly stately pace along the rural roads. Heads turned this way and that as we caught up on what the neighbours had been up to.

Fingers were pointed at glaring mistakes in the fields, either where they had obviously had a blockage on one or two spouts of their grain seeder, or even where the man spreading the fertiliser must have thought his machine had overestimated the capacity and range of his machine, thereby leaving fields striped like football shirts.

These summer trips were not just carried out with the idea of viewing the arable crops and their failings. Whenever a field full of livestock came into sight, comments would take place as to the quality of the stock. This was especially so if they were bought-in store cattle and some grand price rumoured to have been paid. It has long been recognised that one of the symptoms of spring fever in farming has been the paying of a high price for store cattle. Along with an impatience to get onto the land and sow the crops, seeing the grass grow with no livestock to keep it grazed brings on a mild form of madness. Many a bidder lost his natural caution and fine judgement at a spring store cattle sale and went over the odds in his desire to have a field full of good cattle grazing. It goes without saying that this cattle-filled field would either be right in front of the farmhouse or very visible from the roadside.

That was how I first learned about the important role that Fife played in the beef cattle industry. Like many neighbours, the farm that I was brought up on carried a small commercial herd of breeding cattle. In those days, before the introduction of Continental breeds, the most popular was the Irish Blue Grey. Mated with an Aberdeen Angus bull, it would then produce a calf without a great deal of difficulty, mother it well and, without a high level of maintenance, continue to do so for a long number of years.

Fife has never been noted as a powerhouse in pedigree cattle breeding but there have always been a few farms with well-known pedigree herds. In the early days, these were based on Shorthorns or Aberdeen Angus, although throughout the whole of the last century there were also keen followers of the white-faced Hereford breed.

Pedigree cattle breeding continues in Fife. Following the introduction to this country of Continental breeds such as Charolais, Limousin and Simmental in the 1970s, it does so with a wider diversity of breeds. Sadly, and because we were unaware of the importance of keeping a wide gene pool, no one can now provide a definitive view of the original Fife breed of cattle that slipped out of existence in the mid-nineteenth century.

With its black coat and upturned horns, the Fife cow was described as dual-purpose, providing milk as well as producing a beefy type carcase, but its contribution to both these attributes must have been overtaken as farmers turned to newer, more specialised breeds that either produced more milk or ended up with better meat-eating carcases.

Whether from pedigree or commercial herds, the beef calves born and reared in Fife were, in number, always heavily outweighed by those coming in from other parts to be finished or fattened on the residues and by-products of the arable crops grown in the county.

More than half the beef-breeding cows in Scotland live in the Highlands and Islands. Farmers from Fife would make annual pilgrimages to the store sales held in these areas to buy calves for fattening or finishing. With the arrival of trains in the second half of the nineteenth century, store cattle also came in from Ireland after producers there realised that they could get more money by selling them to the UK than by trying to finish them at home. As far as Scotland was concerned, the main port of entry for the first half of the century was the Merklands market in Glasgow. A typical trip would see the cattle loaded in Dublin, then cross the Irish Sea, before being unloaded in Glasgow and driven along the city streets to the Merklands auction.

From the sale ring, the cattle were walked back down the streets to the various railway stations in the city, but the vast majority of the stock was intended for farms in the east of Scotland. Even after their lengthy boat and train trip, often the Irish cattle coming off the trains at local stations, such as Auchtermuchty, Kilconquhar, Luthrie and Mount Melville, still had a long walk to their final destinations. In those early days, it was not unusual for livestock to be herded for distances up to 10 miles by road.

Coming from even further afield were Canadian cattle, first imported in the 1920s into Scotland, an import trade that continued right up until the 1950s. Again, with its reputation as an ideal location for finishing stock, many of those long-legged horned cattle ended up in Fife.

'Ended up' may have been an expression of optimism because even on the final lap of an exhausting journey that lasted days, many of the drovers found the Canadian cattle a 'bit quick' and often they left the drovers far behind, puffing and peching (panting). Any expected sympathy arising from this rodeo-type performance got short shrift from farmers. 'Cattle that cannot be handled need a new cattleman', was the dictum and no doubt the threat was carried out in the early days when staff was plentiful.

Cattle from the wide, open lands of Canada were all horned and often the first job when they arrived at their new home was the removal of these potentially damaging weapons. This gory task often entailed lassoing and tying them to some immoveable object. The vet then used a saw to lop off the offending horns, leaving spouts of blood spattering the walls of the operating shed.

After this performance, farmers got down to fattening the cattle in their courts, where, for the next six months or so, they would be fed a combination of bulk foods such as turnips, silage and hay, and also high protein in the form of dry feed.

In the turnip shed, always strategically placed at the end of the cattle courts, turnips would be sliced up and this would increase the amount that the cattle could eat. In earlier days, before the use of small, two-stroke engines that drove the turnip cutter, the job of breaking up the turnips was done by hand, lobbing them singly into what was called a 'smasher'. This basic piece of equipment consisted of a number of sharp blades shaped into a 'V'. The turnip was placed in the 'V' and a hammer came clattering down, splitting the turnip into slices. There was no such work with silage, which was, in the early days, sliced off in the pit with a large foot-operated knife. Later, this task was made easier by hydraulic loaders working on tractors to prise the silage from the pit.

Because many of the farms in Fife grew arable crops, any part of these crops not destined for market was used as cattle feed. Thus, any potatoes deemed unsuitable for selling as seed or for eating would end up in the turnip shed floor, ready for feeding to the cattle. In years of surplus production, farmers could buy potatoes from the Potato Marketing Board for cattle feed. These would come into the farm covered in a purple vegetable dye to prevent them being used for human consumption. The potatoes may have had a vivid appearance to discourage consumption by humans but the eating quality was not affected one bit – it was, after all, just a vegetable dye.

The other use of this vivid purple dye was in the pre-marital 'blackening', where the would-be bridegroom is stripped naked by his so-called friends and then covered in oil, paint or potato dye, the latter being very popular as it was tremendously difficult to remove.

Care was always to be taken with feeding bigger potatoes as there was a tendency for the cattle to try and swallow them whole, choking in the process. Many farmers sliced their 'brock', or waste potatoes, before feeding them to cattle. Others would leave a short length of rubber hose handy that could be used as an emergency method of clearing the potato-induced choking.

Another bulky food much in favour in the sugar-beet growing era was the feeding of the tops of the plants. The beet itself would be carted off to the factory, leaving the haulm behind. This was quite nutritious and had the bonus of being free; apart, that is, from the cost of the physical effort of collecting the shaws from the beet fields. One danger with sugar-beet tops was that they were always in contact with the ground and therefore often badly contaminated with earth. This feed source was only a pre-Christmas feed option because the shaws did not keep for long after the harvester had been through the crop.

The bulky feeds would then be barrowed or carried in large spale baskets (made from weaving wide shavings of oak together) down the gangway or raised platform between the courts to cattle waiting with their heads through the feed barriers. On cold, frosty mornings, the cattle's hot breath would condense and form a sweet-smelling fug as the 'cattler', as he was named, tipped the food into the deep troughs. This daily routine was carried out seven days per week, and as creatures of habit, the cattle would soon bellow at any slight delay in the performance.

Farmers could hear from their bedroom windows if the cattleman was slightly late in getting to work on a Sunday morning, but when it came to the farmer's weekend on, the men could likewise hear if he was slightly tardy in going about his own tasks. One farmer prone to overindulging in drinking was reputed to have string tied to his big toe as he lay in bed with the other end hanging out the window; the foreman's job was to give this a gentle tug on mornings when his master did not appear.

No sooner had the bulk diet been completed than the feeding of the meal started. These dry feeds had all been mixed to achieve the right balance for each group of cattle. The younger stock would get more protein to help their bones and muscles grow; the older ones heading for the abattoirs would receive more carbohydrates.

Feed would be carried in a sack from the loft and trickled out along the troughs as the cattle gently pushed and shoved their neighbours to get a bigger share of the goodies. The meal itself had been mixed on the loft floor. Generally, the bulky ingredients were tipped out first. Large sacks of dried beet pulp from the sugar-beet factory in Cupar were a staple of most feed mixes. Imported dried flaked maize was another regular in the loft store and a bag or two often added to the heap. Then, with more delicacy as they were expensive, came the high-protein goods with exotic names revealing their origins in far-off lands: Egyptian cotton cake, Palm nut oil, Linseed cake, Molassine meal.

With the heap of feed peaking out, a few farmers added a sprinkler of special ingredients in the form of vitamins with special attributes, or even in the early 1950s, two compounds designed to reduce the potency of the male hormones: stilbestrol and hexestrol. The latter came with claims as to how much extra beef they could put on to the cattle, but also with warnings over what they might do to human fertility, a fact missed by our old cattleman, who sprinkled the powder on with his hands. This mixture was turned over to another part of the loft floor and then back again to ensure a complete mix of all the ingredients. The final part of this dusty job was bagging it up for the days and weeks ahead. Later, the task was taken over by on-farm feed mills. Nowadays most of the dry feed for cattle comes, ready-to-use, in bulk lorries prior to being delivered to the cattle in specialised feed boxes running alongside the feed troughs.

The feeding did not stop there – it was essential to have sufficient roughage in the diet of the cattle. Hay or oat straw would be fed last diet of the morning and last diet at night, leaving the cattle ruminating or chewing the cud until their next meal. Bales would be carried down the gangways, strings cut and the hay and straw fluffed out in front of the waiting heads.

One development in the 1960s was the introduction of barley beef cattle, where the animals were raised on a feed regime of pure barley. This was promoted by Dr Reg Preston at the Rowatt Research Institute in Aberdeen and it was reputed to put beef on more quickly than through more traditional methods. However, the intensive single diet also brought about a great deal of liver problems which reduced some of the initial enthusiasm for the system in Fife.

With all the feed going in one end of the cattle, it is inevitable that a great deal comes out the other end, so the straw-bedded courts needed regular attention. Bunches or bales of new straw were thrown in among the milling cattle, the strings cut and the new bedding scattered around. Anyone at this task always looked out lest they suffer a push or a playful toss of the head by a 500-kilo beast. Depending on cattle numbers and the parsimony of the farmer, this bedding job was done on a daily basis, or it would be stretched out to just once per week.

As the feeding regime progressed, some of the cattle would be considered ready to go to market. In the early days of the last century, the cattle would often be up to 3 years old and tipping the scales at upwards of a ton in weight, but as the years rolled by, the public and therefore the butchers became less keen on buying carcases covered in deep layers of fat which they could not sell.

As long ago as 1954, Anstruther Branch of the National Farmers Union expressed their concern over the down-grading of heavier cattle. Frank Roger, of Kenly Green, St Andrews stated it was obvious that the production of cattle of 15 hundredweight (750 kilos) was not encour-aged. The bigger cattle were being out-graded and this caused a great upset in the East Neuk of Fife, which had for a long time specialised in producing heavy cattle. Now-adays, the market prefers much lighter cattle and the normal finished weight is around 350 kilos, less than half that demanded fifty years ago.

The marketing of finished beef cattle today is largely conducted by selling direct to major meat processors. These direct deals see the farmer paid on a dead-weight basis. Only about 20% of all finished cattle are sold through the auction system.

Selling direct on a dead-weight basis was only brought in after World War II and it was introduced in the teeth of opposition from the local NFU, who did not believe it would give farmers a fair return compared with the sale of live cattle.

Today large-scale abattoirs have taken most of the trade away from local slaughterhouses. There is still a small inde-pendent unit working in St Andrews, but others, such as the one based in Cupar, have long gone. These premises are heavily regulated and the bureaucratic burden hastened the demise of many of the small, local slaughterhouses.

Up until 1954, the Scottish Office controlled all the fat cattle trade and used the graders mentioned above to decide payment levels to farmers. When Government controls came off the sale of fat cattle, the Unions in Scotland and England combined to set up the Fatstock Marketing Corporation, which would buy all the cattle, sheep and pigs coming onto the market. This was possibly the most ambitious marketing move made by the farming industry but it was bedevilled from the start as farmers expected the new body to take all their production without any great regard for what the customer wanted.

When FMC was set up, it was stated that profits would be paid out as bonus payments, but in the first eight years of operation, there was only one year when a bonus was paid. The FMC also incurred the wrath of sheep producers by entering into import agreements on New Zealand lamb to keep the processing factories going. Finally, the company was sold to Hillsdown Holdings, thus ending most farmers' participation in meat processing. Within FMC, the problem was a lack of business management and, typical of many farmer-owned businesses, an attempt to be all things to all men.

Chapter 17

Dairy

IN the early days of the last century most farms in Fife had one or two dairy cows to supply their own and their farm workers' needs. It was only with the influx of dairying farmers from the West of Scotland in the early years of the last century that milk production began to be seen as a commercial farm enterprise. Unlike almost every other aspect of animal production and definitely different from all crop production interests, dairying produced an all-year-round regular income. For some in the hungry 1930s, it was the dairy that saved a number of farms from going under.

The regular payment was looked on enviously by those who relied on a good harvest, followed by a grain cheque once a year; to get their own back, non-dairymen called dairy farmers 'teat pullers'. But the dairy farmer's retort would be, 'We have to have our faces pressed against the backside of cows every day for twelve months of the year in order to get the monthly milk cheque.' Many farmers to this day openly state they could not live without the monthly milk cheque.

The Scottish dairy industry has always been based in the West of Scotland. A combination of summers where the rain helped the grass grow, followed by mild winters was ideal as far as milk production was concerned. To this day, the vast majority of dairy farming is carried out in the counties of Ayrshire and Dumfries & Galloway. However, it is also fair to point out that most of the cropping options, such as growing and harvesting large acreages of grain, were ruled out because of the West Coast climate. It was possibly also the range of options that farming in the East of Scotland could provide that brought droves of farmers from the wetter side of the country, possibly also the fact that farmers in the East had become gentrified and in times of hardship sank into debt and allowed their hardworking West Coast colleagues to take over the farms.

The saying 'Go east for a farm and west for a wife' is still in common parlance. It indicates that while the land is sweeter in the east of the country, females brought up in the west are harder working. For those looking for a West Coast wife, the qualification was reputed to be that the young lady had to have wellie marks on her legs – the sign of a good worker.

The move across country in the early part of the last century must rank as one of the biggest migrations of farmers in this country. Pioneers came through and then sent back word for neighbours and relations to join them. In the early 1950s in Fife, it was reckoned that more than half the farmers were only two or three generations away from their West-of-Scotland roots.

Many from the dairying West brought their own milk cows through on a specially commissioned goods train. Having milked the cows on the old farm in the morning, they carried out the evening milking on the new unit.

In those early days, most milk not required locally was put in 10-gallon (45-litre) churns onto trains heading for the cities. Prior to this, the majority of the cities had town dairies, the city fathers having been persuaded that drinking fresh milk would prevent bone-crippling diseases such as rickets. Many town-based dairies were quite small, with only a few cows kept full-time in byres, providing fresh milk to two or three local streets. By the early 1960s, there were only a dozen or so such urban milk production units operating in Scotland.

Back in the country, many of the first generation of dairy farmers who had come from the West also made butter, which was either sold to travelling vans or again went by rail to nearby towns. The records of one East Neuk farm in 1901 show that skimmed milk was worth 3d per gallon, or 0.3 pence per litre. At the same time, farm-made butter fetched 1/1d, or 5 pence per pound, all of this produced by a hand churn that the farmer purchased at his farm ingoing for £1 17/6d, or £1.88.

Milking on farms with dairy enterprises was carried out by hand twice a day. This work was undertaken by the farmer's wife or by female workers. If the milker was a beginner, an older, more experienced team would check that the cow had been properly milked, or stripped out.

If the milk was to be made into butter it would be sieved into large pottery basins, where it was left until the cream came to the top. This was then 'skimmed' off, either by hand or by using a separator. The cream was put into the wooden hand churn and rotated until it turned to butter which was then kept in the cool of the pantry until a passing grocery van collected it. Often this deal would be on a barter basis that helped pay for groceries for the farmhouse.

For those supplying the raw milk market, the first job was to cool the milk. Warm milk breeds bugs far more quickly than cold milk. This is still a major factor as milk is a staple food and often fed to the most vulnerable groups in our society: the very young and the very elderly. Right at the beginning of the century, the scale of disease in the dairy herd was recognised when, in 1901, meat inspection of carcases of many dairy cows showed them to be tubercular. Tests on live animals confirmed the extent of the disease in milk-producing herds.

Progress in eliminating the disease from herds was painfully slow. By 1935, out of a total of 8,202, only 401 milk-producing farms in Scotland were Tuberculin Tested – somewhat fewer than 5% of all milk producers. By 1941, there were reports of increasing numbers being infected with tuberculosis. Even by that time, only one-quarter of all milk producers had registered under the scheme.

Inevitably when this was raised at a meeting of the Fife Education Committee, one worthy stated he had drunk unregistered milk all his life and that he could see no problem. However, the Government introduced a TB eradication programme. Farmers supporting the scheme in those days had to ensure their cattle were not contaminated by neighbours not in the scheme. Because of this, many double fences were erected. Again, contributing to the law of unintended consequences, the erection of these double fences was reckoned to be a major factor in the reduction in support for the Fife Foxhounds – few of the followers could jump the double barriers.

The eradication scheme proved successful and TB ceased to exist in Fife on 8 March 1954. However, more recently, there have been isolated cases in Scotland with imported cattle.

Realising that bugs breed in warm milk, producers soon installed milk coolers, where milk ran over pipes or rippled boards filled with cold water. After this process, the milk was transferred to churns owned by the farmer. Despite weighing over a hundredweight (50 kilos), these were often hand-loaded onto the collection lorry or farm cart. Only a few farm dairies had loading banks, where the churns could be expertly rolled on their lower rims and then onto the lorry.

These collection lorries would then be seen trundling along the country roads, taking dozens of churns to the creameries to be made into butter or cheese, or – more likely in the east of Scotland – to the milk bottlers and retailers.

Early in the milk-churn era, such were the numbers of farmers in the dairy business that the common practice was for the lorry to collect at six or eight farms within a 2-mile radius. The situation is vastly different today, with only around thirty dairy farmers in the whole of Fife and some 2,500 in the whole of Scotland.

Just as there were very many producers, so too was there a large number of buyers of milk. In rural areas before milk churns were commonly used, milkmen with their own horses and carts would travel round communities selling their produce. Andrew Peddie of Coal Farm, Anstruther recalled there being a large milk tank in the cart. The farm workers' wives and others living in the country hamlets would arrive with their jugs and pails to be filled from a tap on the bottom of the tank. One of those carrying out this rural trade in North Fife went by the name of Watery Willie, on account of his alleged practice of adding water to the milk.

After the practice of customers bringing their own utensils for milk went out of fashion, there was a move to the 1-pint (0.4-litre) glass bottle. Schools were provided with one-third pint bottles as part of a Government sub-sidised scheme which was only stopped in the 1970s: by a politician who later went on to be Prime Minister – Margaret Thatcher.

In the late 1950s, Mr Lohoar of Balrymonth, St Andrews proposed the setting-up of a milk collection service that would use a new tetra pack system of non-returnable cartons. He envisaged costs of 1/2d per gallon, or 1.4 pence per litre, with some 2,400 gallons, or 9,100 litres, being handled daily for the 38,000 population of North-East Fife, but this suggestion for co-operation in milk supply was not taken up.

Back in earlier days, local co-operatives such as Dysart Co-op were major purchasers of milk, but there were also big dairy companies making milk products such as cheese and butter. These included the Annandale Dairy company, which had a distribution base in Dundee.

On the farm, hand milking was the main constraint against increasing cow numbers and thereby building up the business. Normally, a hand milker could only milk up to 10 cows, twice daily. Despite large staff numbers in the early days of the century, the numbers able to milk cows limited increased production.

While Scottish engineers were to the forefront in making the first milking machines in the 1890s, there was no quick conversion to these early devices. The 1897 Highland Show demonstrated five milk machines, following the discovery by Dr Alexander Shields of Glasgow of the first pulsating milking machine. However, sales of the waterpower milk machine, two operated by steam and two, whose wheels were turned by small stationary engines, were all slow. Often producers discarded them because they were erratic, especially in keeping essential vacuum pressure at a level pitch. Many producers who tried the early machines reported higher levels of mastitis – an inflammation of the udder – brought about by faulty and variable suction in the machines.

The big change came, as did so many others, with World War II. Before the war started, it was reckoned only about 10% of all milking was done by machine, but by the 1960s, hand milking had ceased.

In line with the Scottish situation, most commercial Fife

dairying farmers did not make the move to machine milking until the 1940s or 1950s. Robert Mitchell of Drumdreel Farm, Strathmiglo installed a vacuum pipeline down his byres in 1952 to allow him to cast aside his milking stools and pails, thus moving into mechanical milking. Connection to this pulsating vacuum through rubber liners fixed over the teats replicated the gentle squeezing and releasing of human fingers used in hand milking.

The introduction of the milking machine also saw the demise of the old byres, where cows were tethered by the neck, the milker working her or his passage down the row of cattle. In came milking parlours, where the cows were brought into a building specially designed for efficient milking. One style of parlour saw the cattle line up, herringbone fashion, down either side of a pit, where the milker worked alternate sides as the cows came in.

Long before milk parlours came into fashion in Fife, Lord Cochrane of Cults, by Cupar introduced a system of outside moveable milking points, invented by Arthur Hosier, who farmed extensively on the Wiltshire Downs. Use of these outside milking bails meant the cows never needed to come into the farm steading. Lord Cochrane saw a system that did not require fixed farm buildings, where a portable building was used twice a day for milking. The only connection with the farm base was a cart going round the field to bring home the milk.

By 1947, Lord Cochrane could claim twenty years experience of the system on his farm. He said that his herd of 28 Ayrshire cattle had never been kept inside, despite six weeks of severe weather, including temperatures of 40°F (4.4°C) of frost. Even under these conditions, the cows yielded more than 3 gallons (13 litres) per day. He also claimed they were healthier, being free from mastitis and contagious abortion. His herd was the first in Fife to be attested free from TB. However, apart from this single experiment, the Hosier system never caught on in Fife or the rest of Scotland.

In 1957, the introduction of bulk tanks for holding milk on the farm came about, although it would be almost two decades later when the last milk churns were relegated to ornamental use. Use of bulk road tankers for milk collection started in Wigtownshire. It was not until 1967 that this collection system came to North-East Fife after a group of six large-scale dairy farmers co-operated in promising to supply almost 2,000 gallons daily for collection.

In the latter years of World War I, as with many other basic food commodities, the price of milk was Government controlled. However, it was discovered that several Fife farmers had been taking advantage of the shortage of milk and were illegally getting higher prices by selling to unscrupulous dealers. Possibly because some of their members had been implicated in this scam, no action was taken by the local NFU.

A year or two of post-war peace brought increased production and more significantly for the home-based dairy industry, a resumption of butter and cheese imports. However, a rapid slump in price for all dairy produce soon followed. By 1930, the financial position for the whole of the farming industry was desperate. Dairying was no exception and cases of cows being left un-milked were reported locally.

Against this background, in 1931 the government of the day introduced the Agricultural Marketing Act. This major piece of legislation aimed at improving the farming industry, included the setting-up of four Milk Marketing Boards covering the UK mainland.

Of the three MMBs in Scotland, the Scottish Milk

Marketing board was the dominant player. The other two, the North of Scotland Board and the Aberdeen Milk Marketing Board, both operated in the small geographical areas suggested by their names. These Boards were all given monopolistic powers of purchasing all milk – unless the producer was also retailing. They also worked with a pool price for milk right up until their demise in 1994. Part of the monopolistic approach to milk purchasing was to allow surpluses to be taken off the market during the peak summer production period and to be made into cheese, butter or even dried milk. This caused great opposition from the East of Scotland when the Board set up, as it was argued that in this side of the country most milk was needed for direct sales. Less than 20% of all Scottish milk came from the eastern counties, yet almost half the population was based on this side of Scotland.

At the inquiry to set up the Milk Boards, it was argued that producers in the east of Scotland supplying the raw milk market for twelve months of the year had higher costs than those in the West, who produced a lot of milk off-grass in summer, destined for butter and cheese makers. The market for processed dairy products has always been vulnerable to cheap imported produce, so the east of Scotland dairy farmers felt that by setting up the Scotland-wide Milk Board, they were caught up with those producing a lower-value product.

After the introduction of the Milk Board, Fife Milk Producers Association put in a bid for an extra 2d per gallon (0.2p per litre) on the pool price to compensate for their year-round supply. Far from ever being given this financial advantage, all dairy farmers were expected to contribute through a levy towards the costs of purchasing creameries and cheese-making factories. As none of these premises were in the east side of the country, revolt was well and truly in the air: one farmer refusing to pay his levy saw his cows poinded (impounded) and then put up for sale. Echoing the Turra coo incident in Aberdeenshire, where the sheriff poinded a cow from a Turriff farmer opposed to paying National Insurance, at the consequential auction the impounded animals received only a derisory bid of 1 penny from a large audience of dairy farmers silently supporting their colleague.

For most of its life the Scottish Milk Marketing Board operated with the Government, which set a guaranteed price for a Standard Quantity of milk. This system continued up until 1973, when entry into the European Community brought with it milk and butter mountains.

In 1984 quotas on the amount that each farmer could supply were introduced to control dairy production. However, they also brought the full force of European bureaucracy to the farming industry. Initial stages of the quota system were not problem-free and many farmers whose original quotas were miscalculated subsequently received compensation. Then the authorities allowed those who later retired from dairying to keep their quotas. These 'non-producing producers', as they were termed, could make a tidy living from leasing out their quota to those wishing to produce more milk.

Eventually realising this ludicrous situation, European Union politicians decided to take other drastic measures to empty their costly stores. While a lot of the butter deemed low-grade went to Russia, local charities were also offered great quantities. Queues formed at local distribution points as butter was sold off for only a few pence per pound. One small boy told a reporter that the butter made excellent oil for his bike chain.

Returning to the arrival of the Scottish Milk Marketing Board, there is no doubt that its original remit of controlling all milk-processing plants caused the demise of the home cheese-making industry and also the tradition of making butter down on the farm. However, as the years rolled on, the SMMB provided a secure price base for all milk produced and initiated innovative programmes aimed at increasing the quality and quantity of milk. One project saw volunteer dairy farmers receive a monthly visit from a milk recorder. Most of the recorders were female and they would weigh each individual cow's production, morning and night.

This helped build up a picture of the more productive animals and the best bulls to use. Milk records also ensured the all-important butter fat levels in the milk were maintained. Falling below the 3.6% butter-fat level would incur a financial penalty on the milk cheque.

The SMMB also promoted the use of artificial insemination. A development that originated in Communist Russia and was later adopted by Western countries allowed producers access to the best breeding bulls without the expense and the potential danger of handling dairy breed bulls.

In the early days, the semen was 'fresh', having been drawn from the donor bull and then quickly transported to farms, where a vet would carry out the insemination procedure. From that situation in the 1950s, the AI process has progressed to one where the semen is kept frozen in liquid nitrogen bottles until needed; the application now being carried out by a suitably trained and experienced dairyman.

The SMMB also created a network of creameries and collection points throughout the area that it worked. With less than 4% of its total milk throughput coming from Fife, there was only one creamery in the area. This was located in Glenrothes and it came into operation in 1958. Its main role was as a collection centre, but it also dried surplus milk into powder. At its peak it handled some 27,000 gallons per day and more than twenty employees worked there.

In 1994, after sixty years in operation, in a drive towards free markets, the Government dismantled the SMMB and its English and Northern Scottish counterparts. Just as many fought against the arrival of Milk Boards, a similar number bemoaned their demise.

Over the century, the main dairy breeds have changed. Initially, the favourite was the Dairy Shorthorn, with its roots in Cumberland. This breed had the ability not only to give relatively good milk yields but it also left calves that could be fattened up for the beef market. Then the smaller brown-and-white Ayrshire dairy cow, which ate less and produced more milk, found increasing favour far beyond its native county. No doubt one of the driving forces in this change was that many farmers brought their own milk cows with them from the West.

However, the sale of Ayrshire bull calves from this specialist dairy breed never produced any sizeable amounts of cash. In those days, many would be taken to the local market in the boot of the farmer's car, restrained within an old jute bag. The farmer's fervent hope was that the unfortunate animal, with only the head exposed out of the bag, would not empty its bowels in fear during the trip.

Nowadays, all dairy cattle are de-horned soon after birth by burning out the growth points, but previous generations of dairy farmers had to deal with horned animals and would often be hurt if a cow gave a toss of her head at a vulnerable moment. Bedding cattle courts was a dangerous occupation – you had to ensure you were not on the point of a very sharp horn.

Traditionally, the Ayrshire breed had horns that swept forward and then upward, but occasionally, horns would deviate from this pattern. Some breeders interested in showing cattle put special covers over the horns to train them into the conventional shape. Just as today sees youngsters with tooth braces aimed at straightening teeth, so the cows wore training covers that could be tensioned to pull the horns together. The correctness of horns helped ensure that it was easy to tie the cows up in the byre. Also, wayward horns could, and often did, injure other cows. For the few dairy farmers interested in showing stock, the shape of the horns was important, even if there was no correlation between this factor and the all-important milk producing ability of the cow.

In 1915, the first black-and-white Friesians came into the country from Holland, their native land, but the main importation of the breed arrived later from South Africa. With many Fife farmers having their roots in the west of Scotland, this new, higher milk-yielding breed took a long time to establish itself in the area. Then, in the mid-1970s, an importation of Holstein Friesians came from Europe. These high-yielding milk cows were described by one Fife dairy farmer, who counted on selling his bull calves for beef breeding and who could not believe anyone could put muscles onto the specialist Holstein milk-producing frame, as, 'just washing the whole country with white-coloured water.' This jaundiced view did not prevail, however, and nowadays the majority of dairy herds in Fife are based on Holstein Friesians.

Over the years the number of farmers with dairy cows in Fife has dwindled from more than 1,000 in 1900 to less than 30 today. The arrival of machine milking encouraged larger and larger herds. While more than 1,000 farms may have kept at least one dairy cow in 1900, now there are just 24 commercial herds in Fife. This reduction does not automatically mean a reduction in milk produced, though, as the average size of the remaining herds is now well over

100 cows. In addition, yield has more than quadrupled in the past 100 years, top cows producing in excess of 10,000 litres in a lactation.

So far, Fife has no robotic milking machines: one or two have been installed, but have been taken out by the owners, who claim they are not as trouble-free as they would wish. However, there are a number of these self-milking machines in the rest of Scotland, where cows decide themselves when and how often they are milked. Sensors, laser rays and computer software operate to locate the rubber milking cups on the four teats of the animal and then carry out the milking without human help. Whether this anonymous, computer-based system is preferred by the cow rather than milking with cold fingers is impossible to say.

Chapter 18

Sheep

IT was a wet Sunday afternoon in January and the choice was between school homework and going with my father to feed the sheep. After extracting a half-promise from my mother that she would write my essay, I departed happily in the wake of my father. Wellies and oilskins were pulled on, as in those days it always rained on Sunday afternoons in winter. Really, there was no choice in the matter. It was the shepherd's weekend off and feeding fattening sheep with turnips used to be a two-man, or in this case, man-and-boy, job.

As the turnips had been lifted they were pitted in small, straw-covered clamps around the field. There, they would remain in preparation for the arrival of fattening lambs. Fresh from the sale ring or the grass fields, the lambs would nibble away at the remains of the crop while being assured lots of freshly sliced turnip as the bulk of their diet. My job was to feed the turnip slicer or cutter, while my father lifted the heavy baskets or skulls of sliced turnip and emptied them into deep wooden troughs. The latter was always made precarious by a combination of slippery mud underfoot and having to deal with a bunch of hungry sheep that butted and shoved their provider of food.

While my father was coping with these dangers, I tried to pick a slightly less muddy turnip finger from the next basket. Sunday lunch may have been eaten, but this was fresh, raw turnip and anyone who has tried it will swear it tastes superb. There are still few memories more tangible than the raw turnip finger.

Less memorable was the moving of sheep nets as the whole operation of fattening lambs was based on fencing them off to clear up a small area of field before allowing them onto the next break. This system, called folding, ensured a fresh supply of food, but it entailed moving wire netting and wooden posts which were erected every 5 yards or so along the net. Again, this was a tricky job if the going underfoot was as wet and slippery as it always seemed to be. There were always a few language-learning lessons when the shepherd slipped on the mud with a pile of posts on his shoulder. This folding system operates to the present day except fences are now often electrified and relatively easy to move.

Some sheep do not like being penned in; it seems to be against their natural instincts. Old shepherds used to maintain there was only one road for a sheep that enjoyed showing off a talent for jumping fences and that was down the road to the slaughterhouse. Sometimes, especially if the slopes in the field helped the would-be escapee, another net was erected on top of the original one.

The arrival of the electric fence may have been reckoned to be the ultimate deterrent, with a small electric shock passing through any living being that stuck its nose against the wire; however, sheep soon found they could creep under the net without any shock being felt. Round one to the sheep! A good soaking in the sheep dipper for any that escaped would follow; next time it tried to escape, the electricity tingled through the wet fleece. Round two to the shepherd!

Nowadays fewer turnip crops are grown specifically for finishing sheep and the tidying up of the sugar-beet crop through eating the discarded tops is just a memory. Nevertheless, there is a wide range of other crops that still help fatten the season's lambs. After the marketable part has been harvested, crops of broccoli or cauliflower are often tidied by flocks of fattening sheep. In addition to grazing off the crop, another dry meal is often fed, especially when the farmer wishes to finish the lambs off more quickly. With rain or sleet pinging down on the whole operation and mud underfoot, that is how lambs have been traditionally fattened in many of the farms in Fife.

The ownership of fattening sheep may have shifted from the individual farmer or owner to that of a grazier taking fields all around the area, however. Graziers pay on a headage basis and such is the propensity of sheep to keel over and die, they pay on the number of sheep to come off their winter keep, not the numbers they put on.

The taking of grazing has existed for a long time and early in the century, Frank Roger at Balgove was asked to graze his father's sheep on the golf links at St Andrews. These are the same courses that nowadays feature as world-class centres of golf, home to the Open championships and other major tournaments. At night, the sheep were put in fields across from the now long-gone railway line and then herded onto the links during the day. They were then allowed to graze right up to the Martyrs' monument, in the town of St Andrews itself. The only stipulation the herd boy received was to ensure the sheep did not get into any of the house gardens.

Taking a wider perspective over the past 200 years, the sheep industry in Scotland has been marked by geographical divisions. Breeding sheep have lived and bred on the hills, while their offspring come down onto the flatter arable areas to be fattened for the market. Fife has therefore been a fattening, or as the modern terminology goes, a finishing area. That is a generalisation and within any sweeping statement there are exceptions as some farmers produce lambs for the early market and others work successfully with commercial or pedigree flocks.

There are fewer breeding flocks today than was the case half a century ago. My first 'grown-up' job was to drive the farm shepherd up to the top ground so that he could see our 150 or so Half Bred ewes. The driving was necessary – he had originally been taken on as a horseman and never learned to control mechanical horsepower, but the good bit of the job was that I learned to drive. Less enjoyable was that the old shepherd smoked Woodbine cigarettes. Although I did not know it at the time, the ever-present pungent smog in the Land Rover was a dangerous one.

I was also called into duty when lambing came around – my hands were, in those days, quite narrow. If a ewe was in difficulty then she would be held down by the shepherd and I was given instruction in how to help the performance. Sheep have a reputation for being awkward customers and often they start at an early age; that is when they are being born. Some try to come out backwards instead of the normal head and front feet first; sometimes heads tend to appear ahead of the rest of the body and they have to be manoeuvred so the front legs can be caught. One of the more difficult tasks is trying to push back in something that is determined to come out.

Even after the lamb did pop out, the work was only half-done as air tubes were cleared, if need be, and the young lamb bodies wiped dry with tufts of grass or even an old jute sack. Any lambs that were very weak were put into the back of the Land Rover and hopefully the mother followed us back to the farm, where lambing pens were made of straw bales. A few – and mother insisted it was a few – ended up in the farm kitchen, in a wicker basket laid in front of the fireplace. There, they got some heat in their bodies, and soon they were up on their feet and wandering about.

There were always the orphan lambs. Despite vigorous efforts to keep the mothers alive, every farm with breeding sheep had to devise systems to cope with orphans. One trick was to try and foster these lambs onto another ewe, especially if she had lost her own offspring. If a lamb died, it could be quickly skinned and the skin placed on top of an orphan. Provided there were not too many human smells on the second skin, often the foster mother ewe seemed content with this arrangement.

Nowadays there are machines that dispense substitute milk to orphans, thus removing from the shepherd's busy lambing time the laborious, though very enjoyable, task of feeding lambs from bottles.

Later on, past the stage where the new crop of lambs raced about the field in packs, the day of the 'knife' came, when all the male lambs were castrated. The shepherd used a knife to open the sac in which the testicles were located and then pulled them out using the serrated teeth on the other end of the knife. Some of the older-school of shepherds used their own teeth to pull out the testicles, and with an expert flick, spat them out into a nearby bucket. They then made testicle soup, although I do not think they called it that!

The next big date in the shepherd's calendar was the shearing. It was never the big social occasion that it can be in the hills where there are big flocks of breeding sheep. Often the shepherd was left to get on with it, possibly with a boy to help catch the sheep, then clean off the dags, or dirt, from the fleeces and roll them up. The other job for the boy was to use the tar brush whenever the shears slipped and nicked the skin of the sheep. Fleeces were then thrown into a large jute sack called a sheet. These belonged to the company buying the wool. For the past forty years, the British Wool Marketing Board has had the virtual monopoly on this trade, but earlier, in the 1920s the Scottish Wool Growers was formed to buy wool.

September is a busy time for farms with sheep on them. That is when, if they have their own flocks, the ewes and lambs are split, with the latter going off down the fattening road. For farms that buy in lambs for fattening, large livestock floats arrive with three decks full of lambs from one of the big sales. Previously, the main route by which sheep were transported to their finishing quarters was by the railways. Wagons full of lambs would arrive at local stations and they would then have to be driven along the roads to their new homes.

Many of the memories of those days relate to the shepherding of these lambs. If they were straight off the hill and had never seen a fence or dyke before without having the urge to clear the obstacle, then droving was difficult. Passing through villages was also a headache as many of the gardens at the front of the cottages were not fenced. Many a tasty flower was munched by lambs brought up on a rather more sparse diet of heather and rough grass.

Nowadays farmers do not fatten lambs, they finish them. This is an acceptance that the fatty sheep carcase in demand early last century is no longer needed for today's consumer. More succinctly, right up until the 1960s, a fat lamb was a fat lamb. Entries in competitions for the best pen of three fat (note 'fat') lambs in competitions would see three lambs all over 200 lb, or 95 kilos each. Today's champions weigh in about 55 lb (25 kilos). One trencherman notes, 'Where you used to eat one lamb chop, now you need two.'

Despite this major shift in emphasis, the big debate in the sheep sector since the end of the war right up until the early 1990s related to the grading-out of the lambs. In the early days of the century, the eye of a butcher buying from the auction ring was the sole judge. When his purchase was too fat, he would make more fatty mutton pies. However, in the 1930s, when guaranteed prices came into the meat sector, standards had to be established and adhered to. Thus the livestock grader came onto the scene. Initially there was a team of three grading out the sheep at the market: the farmer, the grader and the auctioneer. This was too complex and soon it was left to the grader whose job was to accept those that were neither too fat, nor too lean.

It is alleged that on a day when there were plenty of sheep put forward, the standards were high; when there were few sheep and the market was short, the standard dropped a notch or two. Just as football fans have doubts about the parentage of referees, sheep farmers had similar worries about those grading their sheep. In one case, a local farmer presented a pen of 40 lambs at Cupar market only to have 39 of them rejected. The next day he took the 39 to Dundee, where the grader (the same person) passed them all. To rub salt into the wound, the grader commented, 'I wish you had taken sheep like that to Cupar yesterday.'

As the years have passed and demand changed, so some sheep breeds have slipped into the background while others have appeared to take their place. Early in the century, one of the dominant breeds was the Oxford Down, but it is now categorised as a rare breed in this country. Until the 1960s, the Border Leicester was popular but since then it has seen a decline in demand. Taking their place are breeds such as the Texel, Beltex and Charollais; all brought over from the Continent and reputed to have meat in the areas of the carcase where the money is.

As every sheep keeper knows, flocks are prone to all manner of diseases and disasters. Old medical books have long lists of sheep diseases such as braxy, pulpy kidney or louping ill, but perhaps the one which has hung around the industry more than most has been sheep scab, highly infectious and spread by a mite. It was recognised by the Government as a dangerous or Notifiable Disease meaning that anyone with sheep with scab had to notify the authorities. In 1917, there was an outbreak at Hilltarvit Farm, Cupar after the farmer bought a ram at Perth mart. A lawsuit followed, with the purchaser claiming costs of dipping his flock. Eventually, the case was thrown out unproven as the buyer had mixed the bought-in tup, or male sheep, with his own stock straight from the market.

As the disease spread through almost all the flocks in the country, the Government made it compulsory to dip sheep twice a year, in an effort to get rid of the mites. This laborious task eventually proved successful, with the reported disappearance of the disease in the early 1950s. However, the double dipping continued until 1959. At every dipping, the local constabulary was required to witness that the sheep were properly dunked through the trough and that they stayed in the dip, with its parasite-killing chemicals, for at least one minute, so it was not an uncommon sight to see the local policeman, stopwatch in hand, observing sheep being put through the dipper. Sometimes, if the police were local, they would put on their waterproof leggings and help. If the police were 'townsers', that is, if they came from an urban background, often they just sat in their vehicles and, allegedly, read the papers.

Following the abandonment of the compulsory dipping scheme, the local NFU discussed the situation. One

member gave the opinion that 'the disease was finished' in
this country unless someone re-introduced it from outside
these islands. His words were prophetic and today, after
previous elimination of the disease, once again sheep flocks
throughout the country are being ravaged.

Chapter 19

Poultry

EARLY in life, I was surprised just how many people were called 'henwives'. I knew my father employed someone to look after the poultry and that we had one of the larger flocks of laying hens in the area.

The 2,000 or so hens we kept, in a combination of battery cages, deep litter units and moveable wooden sheds out in the fields, was in the mid-1950s considered a fair-sized enterprise. As it was for neighbours who went down the same route, the trigger for this enterprise was the lifting of wartime restrictions placed on the availability of animal feed.

Someone had to feed and water the hens; also collect and pack the eggs. While there was a woman to look after them during the week, family labour was roped in at weekends. Looked at only from a conscripted labour point of view, the battle between my brothers and me was based on which section of the poultry enterprise we would look after. The battery cage option was easiest and quickest, especially in winter or wet weather, when you could look out and see some pretty dispirited, soaked hens sheltering from the rain. In summer, however, the option was turned around, both for the battery cages and the deep litter, where it could be dusty. Often, an unsuspecting egg collector could disturb a rat looking for an easy meal in the nest boxes.

Our battery cages were among the first in the area and they provided a sloping floor for a single bird, with water supplied by a little trough running just above a feed box. Underneath the cages, with their wire mesh floors, was a flat surface on which a scraper blade was pulled along daily, removing the hen pen droppings into a large box emptied into the fields as manure every week. So, the job was to ensure that the water troughs were full, there was fresh feed in the trough, the scraper scraped the waste and the eggs were collected. The eggs were placed into so-called 'Keyes' trays that in 1930 were designed for carrying eggs by Martin Keyes in his Maine, USA, factory and survive to this day. Thirty eggs filled a tray and twelve trays filled a box.

Although there were few dirty eggs from the battery hens, the same could not be said for those out in the fields and so, these eggs with a covering of feathers or faeces went through a small egg washer. The eggs moved along two revolving screws, which simultaneously brushed to remove dirt, then the clean egg rolled out. Shortly afterwards, egg washing went out of fashion as it was realised that if it was not properly carried out, the process helped spread bacteria.

The hens in the field were kept in wooden huts with about 100 layers per hut. In order to prevent foxes causing carnage, the bolt holes to the huts were closed each night and then re-opened in the morning. We never lost any hens to predatory foxes, but once came across dozens of hens killed by a weasel.

The field system required moving the huts a few yards onto new ground every week or so and this necessitated moving all the feed troughs and water drinkers at the same time. Feed for outside hens was stored in a metal bin with a sliding lid. Inside were the hen-laying pellets that supplied most nutrients but there was also another more basic, but essential part of the hen's diet: limestone grit. This helped provide the hen with sufficient calcium for all its egg-laying performances and a scattering was added to the daily diet. During World War II, there was at one stage a short-age of this relatively common commodity, resulting in a recommendation in one of the farming papers: 'If grit cannot be procured, then get the bairns, or children, to smash up broken dishes to use as a substitute.'

Eggs laid in the field system were gathered in wire-framed baskets or in the more traditional wicker trugs. Care was required in collection as any broken eggs could result in a whole basket having to go through the washer. We did, however, demonstrate the laws of gravity could, for a short period, be beaten by centrifugal force – we were able to show amazed young friends how to swing a full basket of eggs above our heads. This performance was carried out only when adults were not present, to prevent their blood pressure rising!

The third system of keeping hens involved deep litter. Relatively easy to create using existing farm lofts or by building specialist sheds, this system originated in the US. It was completed by covering the floor with a deep layer of peat moss and a few enclosed egg-laying boxes: the theory was that hens would then be able to scratch about, but a great deal depended on the consistency of the litter. If it became too wet, a strong smell of ammonia often pre-vailed; if too dry, there would be a great deal of dust around.

The one big advantage permitted by deep litter and battery cage production systems was the use of artificial light. Back in the 1920s, it was realised that hens could be fooled into producing more eggs if the number of daylight hours was manipulated. Their genetics may have told them to stop laying – the move into shorter hours of daylight came in winter. Traditionally eggs were more expensive in winter than during the summer. Scientists proved that hens could be deceived into continuing to lay through the provision of artificial light, thus part of the routine for looking after inside hens was the setting of time clocks.

All the hens that came onto the farm in the middle of the century were from specialist breeders. Such was the demand for day-old chickens in the immediate post-World War II period that there were literally dozens of small hatcheries producing for this market. The farming and local press would carry whole columns showing sellers of day-old chicks and pullets. Buyers could have a choice of most of the known breeds, such as Rhode Island Reds, Wyandottes, and White and Brown Leghorns, most of which came from the Eastern Seaboard of the US. They, and their crosses, soon pushed out the traditional breeds, such as Scots Dumpys, Silkies, Buff Orpingtons and Scots Greys, now known only to a few poultry enthusiasts. Cross breeding was popular in achieving increased egg production, but it was a number of years before hybrid-laying stock dominated the market. Today, 99.999% of all egg-laying birds are hybrids.

In the early days of the century, all hatched chickens were reared. The farmer may not have wanted, nor needed the vast majority of the cockerels produced, but they were reared and then ended up in the stockpot. Specialist egg producers soon realised they only needed hens and there was a push on to determine the sex of the chickens. In the 1930s, the Japanese discovered the secret, and six specialist chicken sexers from Japan came over to show how to do this – and do it very, very quickly.

In a spectacularly bad piece of timing, they arrived just before World War II broke out. Prior to Japan's involvement in the war, they went to work and in one 16-hour day, a member of the team sexed more than 10,000 day-old chicks. The chicken sexers were interned in 1941, but within months were released at the request of the Poultry Council so they could use their professional skills to help the war effort. There were many letters of protest at the release, but they soon dwindled away and later in the war, the 'Chicken-Sexing Six', as they would be termed in today's tabloids, were repatriated. Despite this innovative process pioneered by the Japanese and quickly adopted here, there was always the odd mistake and each batch of chickens that came to the farm contained a cockerel or two.

Once transported to their new homes in small, hay-lined boxes, the chicks were raised under heated incubators. Often these were kept warm by paraffin heaters. As they grew out of their chicken fluff and gained feathers, they were frequently dosed against a debilitating disease, coccidiosis, which would cause them to lose condition and subsequently stop laying.

For the farmer who did not want the additional problems of raising day-old chicks, the option for building up the laying flock was the purchase of point-of-lay pullets. The theory was that the hard part of the upbringing had been achieved, but this was not always the case as new stock took time to settle down.

No other part of agriculture production has transformed itself so radically in the past century as the poultry industry. Today's laying hen will pop out some 300 eggs annually; two or three times the number her ancestor managed only 100 years ago. At that time, an annual production of just over 100 eggs was considered good going. Incidentally, almost all egg production in the early days came during the summer months.

Today's consumption of white, or chicken meat, now far exceeds that of the more traditional red meat, whether lamb or beef. From a position of eating less than 1 oz (25 g) of poultry meat per week in 1950, the average household now consumes ten times that amount of white meat.

At the start of the twentieth century, most farms would have a few hens scavenging about the yard, picking up grains from the bases of the stacks or even bugs from the middens. There were few commercial flocks and any existing farms were generally on the outskirts of towns where sales of eggs could take place. Eggs would be laid in corners and here, there and everywhere. Farmyard hens scratched about in stackyards and, as harvest approached, in the fields close to the farm steadings. They would perch on implements, in cart sheds, and the flightier ones would even reach the rafters of sheds, where they would roost at night, safe from predators.

It took knowledge of hen habits to discover all the nests but it was important for the woman of the house to do so because eggs were an essential part of barter. The travelling vans might have a range of groceries available to augment food on the farm and it was easier to pay in eggs and butter rather than hard, and sometimes unavailable, cash.

As their natural genes predominated, the hens would occasionally stop laying eggs and start clockin'. This was when the hen became broody and she would sit on a clutch of eggs until they hatched. Generally this was encouraged – a renewal of the hen population was good. In fact, on some farms where the hen population had dwindled, the farmer went out and bought clockin' or broody hens. The accounts book for Drumrack Farm, at Anstruther in 1915, shows that at one point they bought six cluckers (another version of clockin') from a Mrs Anderson of Toldrie. The eggs were mostly fertile; the proverbial cockerel would strut his way around the stackyard, loudly proclaiming his presence and his domain. Although the farmers may not have realised it, these self-same cockerels were often mating with their own progeny.

Regardless of the finer points of in-line breeding, the cockerels strutted about until a feast day or celebration came along, and either he or an old hen or two would be targeted. Once caught, they would have their necks thrawn, or pulled. There was quite a knack to that final act: it had to be achieved with a pull and twist. Removal of the whole head not only left a bloody mess, but also demonstrated an amateur at work. Neck pulling was less brutal than the head being chopped off with an axe, which often left the headless animal fluttering and staggering around.

In general, hens and indeed ducks and geese only played a peripheral role in the operation and economy of the farm in those days.

Until the early 1970s, there were no large indoor specialist units for egg production, or for producing so-called broilers for their meat. Now the poultry sector is firmly divided in two, with meat production carried out in large specialised housing, where batches of chickens can be ready for market less than two months after birth. Egg production is more diverse, with a range of methods, from batteries through to free range. Both are conducted in much larger units than previously, doubtless without any antics, such as swinging pails of eggs around.

The marketing of eggs has also been transformed from the day of farm eggs being bartered for goods from the men in the travelling grocery, butcher and any other vans. Now there are large-scale packhouses, although there is an exception to that statement with the arrival of Farmers' Markets, where eggs move directly from producer to consumer.

The suggestion of co-operation in egg marketing came in 1929 with a proposal for a marketing group in the East Fife area: the idea was that a collection centre be set up to test, grade and pack eggs for members. It was estimated the co-operative would handle around 2,000 dozen eggs per week, and through better marketing, this would help compensate for over-production in the area. However, this ambitious scheme never made it beyond the talking stage. Almost a decade later, in 1937, in order to raise their prices, the Fife Poultry Keepers Association decided to grade their eggs into standard and small sizes.

The next move in egg marketing came in 1941 by which time egg rationing was brought in by the wartime Government. Regulations required all producers with more than 12 hens to be registered. Those registered had to supply all their eggs to the 'pool', but inevitably there was discussion at the NFU meeting as to how the Government could ensure that every egg had been submitted. They sent inspectors to check that records of all egg production were kept.

The same year saw a great deal of agitation from producers for an egg-packing facility in North-East Fife, the nearest one being out of the district at Crossgates. Egg producers were told that a packing station required a minimum annual delivery of 2,500 crates, each with 30 dozen eggs. This would economically justify the £500 needed for equipment, this money mostly being used for grading eggs into different sizes. However, they did not get their packing station until much later, when a small unit was set up in Kingskettle. This operated through the 1950s, but closed when rationalisation saw bigger packing stations come into use.

Chapter 20

Pigs

LIKE many other mixed farms, Logie kept a few pigs, six to be precise, with two each in three adjoining sties. In no way did this small enterprise need any full-time labour; it was left to the orraman to feed the pigs twice daily, in the morning before he started his official work and then after finishing the day's graft.

In the morning, as they grunted and pushed each other out of the way to get their snouts in the cast-iron trough, they would get a wet mixture of pig meal made up largely of barley, with a little fishmeal as protein. Apart from the cost of fishmeal there was also the danger that too much could 'taint' the pig meat. In the evening, the diet moved on to boiled potatoes. These were the spuds rejected during grading of the seed and ware (for human consumption) potatoes. The boiler was a large cast-iron affair with room for a coal fire beneath. Up to 440 lb (200 kilos) of potatoes were emptied into the boiler; after they were judged properly cooked, the whole machine was emptied. My brothers and I splashed in the boiling water, trying to grab a freshly boiled potato before they were fed to the pigs.

That was back in the 1960s and these pigs were among the last of the small lots kept on farms. Thereafter, keeping pigs became a large-scale enterprise, with specialist producers adopting a far more professional and scientific approach to the production of pig meat.

Back in the early days of the last century, pigs were kept on farms to convert the farmhouse scraps and poor-quality grain into meat. Some dairy farms kept a number of pigs and fed them on the whey, or residue of milk after butter and cheese had been made. In the early years of World War II, pig production was drastically cut back, due to a shortage of animal feed. One industry commentator said, rather dramatically, that 'pig producers have got to make sacrifices in order to exterminate the Nazis.' The Government cut the market price to emphasise its determination to reduce pig numbers. A similar move had been made in the latter years of World War I, when the authorities reduced pig numbers through feed restrictions.

Despite the food shortage, as the War wore on there was a gradual increase in the popularity of pig keeping. A fat pig could make around £25, a fair amount of cash when a man's wage was only £3 to £4 per week. Pigs produced in wartime had to be sold to the Ministry of Food and they were taken to one of the local bacon factories. There has traditionally been little pork production in Scotland and most pigs went away at 'bacon' weight, between 220 and 240 lb, or 100 to 110 kilos.

In the war years, forms had to be filled in for the Ministry of Food indicating the numbers of pigs on farms and when those being fattened would be ready for the slaughter-house. However, even in those years of rationing, it was not unknown for local milkmen, or butchers, going around the farms to be able to pick up a nice piece of ham.

Up until the last quarter of the twentieth century, the pig industry was split between those who bred pigs and sold piglets as 10-week old weaners and those who then fattened them up for market. The intermediary in this trade was the local auction mart. In Cupar, there was a small ring dedicated to selling weaner pigs and every week small pigs would be brought, squealing in their unique 'I am being murdered' tones, into this sale ring before heading off to be fattened.

In 1947 the Government urged a doubling in pig-meat production to help feed a hungry nation. Taking this as tacit support for the industry, farmers increased their fattening numbers by running weaners in their cattle courts during the summer months. This made use of the buildings and also of the cattleman, who was pressed into temporary duty as pigman.

The start of specialisation in pig keeping came shortly afterwards and coincided with increased knowledge of pig breeding. With the main fattening food being barley, the ability to convert foods efficiently was an important factor in determining profitability. Boars were tested at central stations for their live-weight gain and only the better ones were used for breeding. From that point on, casual pig production ceased and became a highly specialised enterprise. Fife has never been a large pig-producing county, leaving that honour to Aberdeenshire. Some would say that the domination of the Scottish pig herd in the North-East was due to the proximity of processing factories. Others, especially Fife farmers, reckon it is because grain grown in that northern area allegedly resembles 'pregnant gramophone needles' and is only fit for pig food.

However, in the latter half of the century several large-scale farm-based pig enterprises existed in Fife. Although it never actually got off the ground, potentially the largest pig unit seen in Scotland was one proposed south of Glenrothes. In July 1963, an announcement was made in the name of the Royal Victoria Sausage Company, that it would build a large piggery in Fife. The company was part of a wider group belonging to USA-based Cadco, part financed by George Sanders, a film star nicknamed the 'Cad'. The proposed piggery would cope with 20,000 pigs and the announcement of this scale of development was welcomed by politicians in an era when economic news was often downbeat. Construction work soon started, but within a few months, rumours were circulating that the company involved – which had never before engaged in any building operations – was in trouble. Apart from that, the design of the buildings was totally unsuitable and the effluent system inadequate for 20,000 pigs.

Neither the buildings nor the sewage systems were ever put to the test as the whole business collapsed, leaving many local firms and workers heavily out of pocket. No action was taken against those behind the scheme as they were mainly US-based, but the local MP described one of the main perpetrators as a 'treacherous, lecherous character of the worst possible type.'

In the early half of the last century, most of the pigs were either traditional British breeds, such as the Wessex Saddleback, Berkshire or Large White, or a cross between them. One of the top pig breeders in the early days was Lawson of Carriston, Markinch, who kept pedigree Large Whites and also Tamworths. However, in the early 1950s, an importation of Landrace pigs from Finland soon saw many producers move towards this leaner, faster finishing breed.

Nowadays most commercial pig production is based on pigs sold by large breeding companies, who use hybrids or crosses bred specifically to convert food efficiently. Some 90% of all those hybrids carry some Landrace genes. The Large White has its own niche in pig production, being the most important breed in the world. Currently, the UK exports Large Whites to more than sixty countries around the world. This export trade is no modern phenomenon as, in 1931, some 200 pigs were sent to Russia. In this export batch, the most popular breed was the Large White, followed by the Middle White and then the Berkshire, with a few Tamworths and Saddlebacks making up the consignment.

Currently pig production in Fife is based on a few farms, with the sows often farrowing out of doors in small shelters. At weaning, their piglets will move into large, straw covered sheds, where they will be finished for the bacon market. The see-saw between profitability and loss-making in the pig sector has become legendary and is known as the 'pig cycle'; the only problem for pig producers is knowing precisely at which point they are in that cycle.

Chapter 21

Auction Marts

THE distance between Bell Baxter School and the livestock auction mart in Cupar could be covered in about ten minutes, and that might be achieved even at a schoolboy pace. It was therefore possible, together with several friends of an agricultural inclination, to go down to the mart and watch proceedings during the lunchtime break from lessons.

Entry into a livestock auction sale ring was similar to coming across some strange ritual. The auctioneer, very much in control, was, in the clichéd phrase, the ringmaster. He would quickly assess the stock coming in to be sold, then like all livestock auctioneers until a few years ago, he would next take a few words of instructions from the sellers. With a promotional call of, 'This is one of the best pens today,' or, more ambivalently, 'You won't get many better bargains,' he would attempt to get the bidding started. It was soon apparent that auctioneers were, and are, an optimistic bunch as they invariably started looking for bids right at the top end of the expected range.

A quick dose of reality soon arrived as not a move was made, not a finger raised or even the slightest nod of the head from the potential buyers round the ring. And then, as a note of desperation seemed to creep into the auctioneer's voice, the bidding started.

Not that we as schoolboys could see a great deal. He

would point and say, 'The bidding is with you', but so long as it was not in our direction, we felt safe. Returning to school with a pen of cattle, we reckoned, would be somewhat difficult to explain.

Bidding then slowed down as the auctioneer tried to squeeze the last few pounds out of the sale. Meanwhile, the seller, who never looked anything other than a man whose good stockmanship was being completely undervalued, kept shaking his head. Eventually, after a few pleading last calls, the gavel came down, with the auctioneer briefly announcing not the name of the purchaser, but the name of the farm. Mr Smith would have to pay the cheque for the stock, but the only publicity around the ringside went to the name of his property. This custom continues to this day, with the farmer known by his farm name.

Among the many other finer points of livestock trading by auction the schoolboy observers missed were the decoy bidders. This practice happened less in the normal week-by-week store sales, but came into its own at the top end of pedigree sales or even on the very odd occasion when a complete farm would come under the hammer. Often the real purchaser would use a decoy bidder to throw the opposition off the scent.

Another aspect relating to the selling of livestock by auction and one conducted away from the public gaze, being a deal between buyer and seller is the issue of 'luck pennies'. These were originally a small coin given by the seller to the buyer to seal the deal. The practice supposedly conferred luck on the buyer so that the stock would thrive and not fall ill, but it fell into disrepute as wily or cunning buyers could imply there might be a complaint to the seller about the livestock unless a healthy sum of money was transferred.

The practice was particularly prevalent in the dairy trade

and it was not unknown, if the luck money was insufficient, for the buyer to advise the seller he would never
purchase another beast from him. The luck penny moved
quickly from coin to paper money. One enterprising
fellow soon cottoned onto the fact that the seller coming
out of the ring might not have identified the buyer and this
interloper moved in, asking for 'luck money' and often
getting it before disappearing quickly before the real purchaser came into sight.

'Luck' has operated for most of the past century. Even in
1914, Henry Watson of Drumrack Farm, Anstruther faith-
fully records paying 10/-, or 50 pence, for luck after
buying some £21 worth of store cattle at Anstruther
market. Often one of the main Irish dealers to Cupar
market in the 1940s would give a donkey as luck, thus
taking cash out of the equation. He would frequently see
his offer rejected by buyers fearful of retribution on their
return home with a braying companion. Either through
inflation or increased leverage in obtaining luck, there
have been periods when whole bundles of used notes
changed hands after the official deal was done.

Despite all my youthful diversions at livestock marts, I
could never claim to be either an expert in assessing stock,
nor in the buying and selling of them.

The modern auction system evolved from the trysts and
market fairs, where farmers would bring their livestock to
be traded. In the early days, it was aided by a change in the
taxation system that removed the commission rate from
the deal. Some of these markets, such as the Lammas Mart
in St Andrews, exist to this day, but they have replaced the
dealing in horses, cattle, and to a lesser extent, sheep, with
candy floss, bingo stalls, flashing lights and loud music.

In the days of the drovers, major hubs such as Falkirk
would see thousands of cattle gathered after being herded

hundreds of miles down the drove roads that cut through Scotland, but by 1900 the once famous Falkirk Tryst was dying on its feet, with only 200 cattle on the park. The rapidity of the downward spiral was evidenced by the fact that only twenty years previously some 80,000 sheep were sold at the same location.

It was only with the arrival of the railway train and the ability to move livestock from the breeding grounds in the hills and islands of Scotland onto the softer sweeter lands in the east of Scotland, where cattle and sheep could be fattened, that the auction system took off. In the words of one commentator of the day, 'The Scottish runts off the hills could thrive on the lush grass.'

There were reputed to be more than 200 livestock auction markets in Scotland at the beginning of the twentieth century, but Fife's role in this type of trading was only ever peripheral. This was partly due to the prominence of marts at major railheads such as Perth and Stirling, but also due to the type of livestock farming that took place in Fife.

Most of the big markets were based in the stock breeding and rearing areas of Scotland, or in the major towns and cities. With a large proportion of its land suitable for growing crops, Fife did not qualify as a stock rearing area. Its reputation was based on finishing stock off the by-products of prime arable acres prior to them going off to slaughterhouses or to fatstock sales. But its reputation as a good area for finishing stock also came from the large number of shops belonging to the co-operative movement. Because these premises were located in widely differing communities, from mining villages to market towns, the type of beef required in the shops also varied. The net result was that the Co-op was always a good buyer at the marts in Fife.

As in other parts of the country, the livestock markets in Fife were often based close to the railway stations. Cupar market started in 1860, just a year or two after the railway line came through the town. Previously there had been an informal market without permanent pens or stalls half a mile away in what is now the Fluthers car park. Further south on the same rail line, a market was set up at Ladybank, then another two auction marts were established at Thornton. This last location had the advantage of being based at a rail junction, with lines snaking off east and west. This was a major plus in the days when there was no road transport of stock other than on the hoof. Heading eastwards, the line came through the East Neuk and a mart was established at Anstruther to the west of the town.

Further west in Fife, Dunfermline also had a mart, the only one in Fife to hold regular sales of pedigree stock because it had a considerable trade in dairy cattle. The Dunfermline mart was held in the middle of town and, even in the early days, there was often congestion around the area. It was reported that a well-known auctioneer, Oliver Cumming, was held up on one occasion. Leaning out of his car, he shouted to the traffic policemen, 'Don't you know I am Oliver Cumming?' to which the unimpressed, possibly harassed constable retorted, 'I don't care if you are Oliver Cromwell!'.

Just outside the county in Milnathort, there were two auction companies. Again, both close to the local railway line – one on either side, in fact. Both closed down in the 1960s, the last sale at United Auction being in aid of the local Tory party, and both sites are now covered in houses.

Early sales of fatstock and store in Cupar were conducted fortnightly by a David Whyte. By 1885, the auction business was run by the firm of Walker and Morton, but the name synonymous with selling horses, cattle and sheep in Cupar, Speedies, took over two years later. Alex and Matthew Speedie already had a livestock auction business in Stirling and also operated out of premises in Edinburgh owned by the city corporation. The brothers were among the first to make a full-time occupation out of selling livestock.

Previously, this job was seen as something any auctioneer could do. Furniture one day, cattle the next, used to be the auctioneer's trade, but the last 100 years produced the specialised livestock auctioneer.

In those days, large conurbations regarded it as part of their civic role to hold livestock sales within their city walls and the Speedies also had a base in Glasgow. Other firms leased the corporation premises in Edinburgh and Dundee. However, their purchase of the Cupar business seemed to shift the emphasis of their life as they bought or rented a large acreage of land in North-East Fife and South Perthshire.

One of Matthew Speedie's farms was Kinshaldy, just north of Leuchars. Always considered very much second-class land, it was basically part of the raised beach and as such the land was seen by agriculturalists as nothing more than running sand. However, the shrewd Matthew Speedie realised that the scrubby grassland could be greatly improved by heavy stocking. His company was in the business of dealing in cattle and sheep, and according to one record of the day, they put large numbers of sheep and cattle on the land. Not content with that, Mr Speedie then built one of the finest farm steadings of the day at Kinshaldy. Though the stone had to be carted several miles from the nearest quarry, the wood, the finest red pine, was salvaged from the nearby beach with the wreckage of boats failing to navigate the tricky sand banks at the entrance to the Tay. With its shifting sands and sandbanks, the area around Kinshaldy, south of the Tay estuary, is still reckoned to be one of the more treacherous areas.

Not long after establishing themselves in Cupar mart, Speedie Brothers saw the potential for the market in the East Neuk of Fife and bought over the sale premises in Anstruther belonging to the Fifeshire Auction Company. This area was long renowned for finishing cattle and sheep. There, they ran fatstock and store sales on alternate Fridays. The last weekly sale in Anstruther took place in 1955 before the site was developed into housing.

In its early days of the last century the Ladybank mart was also run by the Fifeshire Auction Company. An example of the scale of operation at Ladybank is shown with the annual turnover in 1910 of £1,400. With sheep trading at approximately £1 per head in those pre-World War I times, there were few big sale days.

Just as he had done in Anstruther, Mr Speedie then took over the Ladybank mart from the Fifeshire Auction Company but he only kept it for a few years before selling, in 1921, to another well-known firm of auctioneers, MacDonald Fraser, who kept it going until 1937.

After the fatstock restrictions introduced prior to World War II were removed in 1954, Ladybank mart closed its doors for the last time, although sales of house contents continued to take place until the early 1970s.

The year 1921 must have been a time of considerable change for Matthew Speedie as the rest of the company also went to MacDonald Fraser, although the Fife-based operations continued to operate under the Speedie Brothers banner. This sop to ensure staff remained with the new company did not work as several 'Speedie's men' left to set up a rival company in Stirling called Live Stock Marts.

It is interesting to note that in 1962 MacDonald Fraser amalgamated with Live Stock Marts to form United Auctions. For a short period, it was therefore possible to state that which had been cast asunder had been joined again. However, history repeated itself with this 1962 amalgamation as several disgruntled staff again went off to set up another auction company. This time it was called the Caledonian mart and it was based in Stirling. The mart not only survives to this day but it flourishes along with United Auctions which also operates in the same area, making Stirling the only town in Scotland with two marts.

In its heyday, Cupar had been one of the busiest markets in Scotland, with a large throughput of store and fat cattle. Old sale particulars show entries of more than 500 head of cattle on the Tuesday market day. One particular Irish cattle dealer, Jack Coulter, used to bring over around 400 head and have a special one-day sale for his cattle.

The Scottish trade was important for Irish farmers: the cheque paid by the Scottish market to the Irish dealer would go through several hands before being cashed. Banks no longer return presented and cleared cheques to their customers but when they did so, it could be seen from the various Irish signatures that the cheque had gone through a number of farmers before finally being cashed.

Many of the cattle came across from Ireland via the big market at Merklands in Glasgow, where teams of auctioneers from all over the country were allocated times in the auction rostrum to sell the Irish cattle just off the boat. The Merklands sale was held on a Monday, but if the cattle needed a recovery period before appearing in the ring, the auction companies would use their own land for holding stock.

These same parks were used as holding areas for cattle

and sheep not sold at any previous sale. They were also useful for ensuring the brought-in cattle were not carrying any disease. Vets were employed by the UK Government to check the health of cattle prior to their departure from Ireland. In the 1930s, the biggest problem with the milk cows that came over was transit fever which, in those days without antibiotics, could not be treated.

The other problem that the vets reported with the export trade in milk cows from Ireland was keeping the street urchins from milking them as they waited to be loaded onto the boats. One health inspector stated, 'The boys from Dublin were quick and whenever the opportunity arose, they would milk the cows into cans which they carried.'

In the busy days in the 1960s, the area around Cupar market was a hive of activity, with livestock floats backing into the unloading pens, buyers and sellers mingling around the mart area. This level of congestion saw parking restrictions brought into place by the Council.

By the 1990s, Cupar was the last surviving market in Fife, but in 1993, in a decision that alienated quite a few regular customers, the board of United Auctions took the decision to close it down. The last-ever livestock auction sale in Cupar, which was also the last in Fife, was the Christmas sale in that year. In a glimpse of yesteryear, it was reported not only was there a large entry of stock for this last event, but a big throng of buyers and sellers also in evidence – or, as one of the former auctioneers ruefully remarked, 'former buyers and sellers.' The county had never featured highly in the auctioneering business and most of the operations were smaller satellites of larger companies based in Perth and Stirling.

Over the past century, the visitor to markets would see a massive difference in the cattle and sheep coming into the

ring. In size the early century cattle would be up to twice
that of those coming through the ring nowadays. They
would also be twice the age and carry the weight on a big
frame. In shape, today's cattle have a leaner, fitter line
compared with the extremely well-covered carcase of
earlier decades. There are fewer horns about nowadays,
either following breeding lines for polled stock, or due to
the removal of the horn buds when they are calves.

In colour, there would be the reds, blacks and greys of
the traditional breeds in the early part of the century. Apart
from Irish cattle, the favoured breeds in the 1940s were
Aberdeen or Hereford-sired stock, the former with its distinctive black coat and the latter having a prominent white
face that can be seen all over the world, including every
cattle-driving Western!

Fife, possibly uniquely north of the Border, had several
pedigree herds of red cattle, Lincolns and Devons, and
these distinctive cattle were always sought after on market
days. To them are now added other textures of Continental breeds: the off-white of the Charolais, the red/brown
of the Limousin, the creamy white/light brown of the
Simmental, or a combination of all of these.

Our ancestors were brought up thinking that the only
beef that was good had a nice thick layer of fat around it.
When cooked, this fat would sizzle and seep into the red
meat, providing the classical roast beef meal. In those days,
meat without fat was reckoned to come from the carcase of
some old cow or ewe. The mental linkage was that the
meat from the working animal without the fat would be
old and therefore tough or stringy, or both. There was also
a tremendous pride among the farming fraternity for big,
really 'big' cattle.

Before direct selling of prime cattle to processors and
abattoirs came into vogue and before graders set standards

and upper-weight limits for the finished carcase, the average weight for finished cattle was between 14 and 15 cwt, or 700 and 750 kilos. At the same time as the heavier types were being sold, several Fife farmers vied with each other to produce the heaviest animal in the fat cattle ring.

A memory from Jimmy Garland, who for a long time served as auctioneer at Speedie Brothers in Cupar, picks up on a bullock from John Bayne of Magus, St Andrews, weighing in at 1 ton. This set the competitive juices going among his farmer colleagues and it was not long before John Small of Vicarsford Farm, Leuchars came forward with a beast weighing 24 cwt or 1,200 kilos when it staggered majestically into the sale ring.

All this came to an end, when butchers realised they were buying lots of expensive fat. They found they were only selling a small percentage of the meat, the rest of the fat heading for the not-too-valuable tallow trade. In the early 1920s, after a feisty fight between sellers and buyers, who up until then brought cattle and sheep into market without weighing them, scales were introduced so that buyers at least knew the weight of stock bought. The Irish sellers fiercely resisted this move, believing the buyer's eye was good enough. With a good bit of power, they boycotted markets with weighing machines. After a period, however, they accepted these were part of the selling system and nowadays all stock goes over the weighing machines before sale.

A critical part of the livestock auction business is canvassing for support, and this was, and still is, an essential part of a successful auctioneer's responsibilities. On non-selling days, he is to be seen around the farms getting orders for forthcoming sales. A good auctioneer knows the farmers and when these farmers like to sell their stock, when they prefer to buy and what kind of sheep or cattle suit their farm.

Valuations

Another essential part of the auctioneer's life was conducting valuations. These could be for a variety of reasons: they may have related to a change in ownership or tenancy with fixed equipment and crop requiring a valuation.

In the early days, this would often be carried out by an auctioneer or arbiter with the experience required to measure up a midden of farmyard manure, or estimate the tonnage of turnips in a field or a store. Little books were carried to help in the calculation of the cubic capacity of a cone or a cube. Then again, from the book a figure for density of the grain, and within a short time, the tonnage and value could all be worked out.

Straw stacks would be measured around their outside and from the geometric formula, the surveyor worked out just how much straw was in the stack. Dung middens were stepped out for length and width, and the overall height estimated. Thereafter it was easy to work out the capacity of the midden, but what was not so easy was deciding just how well rotted, or set, the manure was. If it was strawy and fresh from the cattle courts, there would be one figure for the density, but if it had settled down into a well-rotted state, a different multiplication needed to be done.

While all the responsibility was vested in the auctioneer, he was invariably accompanied by a youngster setting out to learn the trade; to him fell the menial tasks: 'Go up and check the depth of that midden' or 'Count how many fence posts are in that heap.' Faced with the former task, one apprentice of years gone past got up on top of the midden with the steel rod, which was plunged down until it hit hard ground. From the markings on the rod, the depth would then be announced. This time little depth could be found before the rod hit an obstacle. Only when a graip was brought and the top dung removed was it revealed that the midden was also the last resting place of an old horse.

Fields were measured by chain length. Pre-metrication, the chain was 22 yards, or 21 metres, and any auctioneer or valuer worth his salt had a multi-linked chain for measurement. Combined with a set of pins and a willing youth to take one end, the lengths and breadths would be worked out as they moved up and down the field. Odd corners were added or subtracted, the general rule being to try and split the field into a number of geometric shapes: squares, rectangles, triangles, etc., where the area could be calculated by formula.

A great deal of this field measuring took place during the summer months. Potato merchants, who rented fields on a one-crop basis from farmers, wished to ensure they were not paying any more than they needed to for land rented for growing the crop. Such potato rental deals did not include headlands and even in North Fife, 'rock-heads', or rocky non-plantable areas, were subtracted from the rented acreage.

With his knowledge of the livestock industry, valuations of the various cattle and sheep on the farm would be relatively easy for the experienced auctioneer. But then there was the machinery, where experience counted, especially the case when equipment specific to an enterprise, such as potato production, had to be assessed. With a thorough valuation, even small tools and bits and pieces had to be ticked off in the valuer's book.

A valuation book relating to the 1930s shows staithels, the stone bases for stacks valued at 10/- or 50 pence, bothy beds at 2/- or 10 pence, clothes poles at 1/- or 5 pence each, pig sties or crays at 40/- (£2) and the kitchen grates in cottages at 20/- (£1).

Farm sales or roups

One of the highlights and busiest times of the year for auctioneers comes at the term time. In Fife and in many other parts of Scotland this is at Martinmas, or 28 November, the period of the year when it has been the tradition to change over the ownership of farms. Leases and rents are almost always written to operate to the Martinmas term. In England, the changeover date is some 6 months different, coming at Whitsun term. Nowadays, the convention on time has been discarded and farm sales are held at any time of year.

If there is a shift of farmer, owner or tenant, the usual procedure is to hold a farm sale, or roup. For weeks beforehand, the farm staff is engaged in pulling out all the machinery from the depths of the implement sheds, even from the long grass in the stackyard. The cultivators and other working machinery may not be polished up like a second-hand car on a showroom floor, but they are often cleaned up before being lined out in rows on a nearby field. As this work goes on, memories are kindled on seeing bits of machinery long past their working life, and in many cases almost forgotten. Thoughts of yesteryear can even be raised with just a glance at a bash or dent on a well-used piece of equipment, with the worker remembering just how and where that particular incident, possibly even accident, took place.

The general rule is that the implements are laid out in decreasing levels of importance so the largest, most modern tractor or the farm combine takes pride of place. As horsepower levels drop away, the better or more modern implements come next. Soon, the lines are used to lay out sheep netting, fence posts, spare tractor tyres, oil drums and all other paraphernalia linked to the farm.

The laying out of farm gear does not apply to tools, as these are dealt with by handing them out in batches, trays or boxes from the barn door at the start of the sale. In many cases, the original use of a piece of metal has been lost. This was less of a problem in the old days as all machinery was relatively straightforward, but in the more complex world of today, spare parts bought and never used are unearthed and offered for sale at a fraction of the original cost.

Roups always throw up surprises. An item of equipment where the original purpose in life has long fallen foul of the march of time may be bid up to extraordinary amounts, if it becomes fashionable to non-farming buyers. Horse brasses and harness have been through this process. Old potato riddles are currently somewhat in vogue and who knows what else will take the fancy of the non-farmer buyer.

Meanwhile, any cattle and sheep to be sold are also spruced up and split up into even batches. This will help the buyers, whose choice will then be to pay a little extra for the good sheep or cattle, or take a chance in trying a slightly poorer pen at less money.

On the day prior to the sale, the auctioneer will come around and, along with his clerk, will not only label the items but get an idea of the quality of stock and plant for sale. With fewer farms, nowadays there are fewer of these occasions, but Jimmy Garland remembers one term time forty or so years ago having no fewer than 14 roups to be carried out.

In those days, roups would be held any day of the week but this has changed so that the vast majority now take place on a Saturday. This day of the week suits most potential buyers, near or far. On the day itself, the country road is busy from early on with a wide selection of old lorries, pick-up trucks and four-wheel drives with trailers bouncing along behind.

The first part of the sale sees a mêlée around the barn door. While many are just there out of inquisitiveness, others will be neighbours still adhering to the old convention of getting their name on the roup roll. This custom was born in the early days of the century when the outgoing farmer or tenant would often leave the place with very little cash and it was a sort of going-away present to a former neighbour. Unrelated to the roup, another custom saw neighbours often providing a day's ploughing as a welcome to an incoming farmer.

After laying out the conditions of the sale, an often-draughty shed is equipped with a desk, a couple of chairs and nowadays a computer terminal so that all transactions can be recorded. Among the terms of the sale is the all-important settlement on the day, with no delayed payment terms.

Having cleared all the junk from the tool sheds, the auctioneer moves along the lines of machinery followed by a serried caravan of buyers, sellers, plain observers and that unique bunch, 'the tyre kickers'. These people, who also attend local agricultural shows and visit machinery stands, get their name because with never a thought of purchase, they will still, in absent-minded fashion, kick the tractor tyre and talk about buying the machine.

Normally walking quietly at the back is the owner, who, as each piece of machinery heads towards a new home, will feel just another little tear on the mental fabric attaching him to the farm.

For the auctioneer, it is a busy working day. He has to ensure his bidders are prompt with their nods and winks and that they are decisive. Anything less than full attention throughout the sale could mean a missed bid or a bid wrongly taken.

War time at the marts

As food became short in this country during the latter years of World War I, the Government brought in restrictions on farmers selling their stock. The aim was twofold: first, was to ensure the army was fed – an absolute priority in the first major conflagration. Then there was the need to prevent richer members of society from buying scarce food while the poor went hungry.

The policy required all livestock to go through a controlled market, which by implication saw the suspension of the auction system. Graders were appointed to split the cattle up into various payment categories. Both the Anstruther and Cupar branches of the NFU had lengthy debates as to the qualities and attributes of these men, who could make a considerable financial difference to the farmer.

There were allegations that some markets ensured better grades for the stock sent to their premises, and this was only sorted out when the Government passed legislation that required farmers to take stock to the nearest market.

The auction companies were paid a commission on their throughput but there were also complaints about the volume of paperwork and everyone seemed happy when normal trading recommenced in June 1920.

A similar scheme on the marketing of all fatstock operated from the outbreak of World War II in 1939 right through to 1954, by which time normal trading at auction markets resumed.

Chapter 22

War

THE farmhouse in which I was brought up was on two storeys. One obvious delight for any small boy was the banister, which provided a quick and exciting method of descent from upstairs. However, the real and rather spooky attraction of the stair was the cellar below. It was in this extremely dark and rather claustrophobic sanctuary that the family took shelter from the German bombers.

Like almost everyone in reserved occupations such as farming, my father was in the Home Guard and many a night he would be on duty, watching enemy aircraft overhead. Mostly they were headed for Clydebank, where they hoped to bomb the shipyards out of action. In this part of the world the majority of bombs were dumped by planes returning from the west of Scotland. They were dropped on a random basis mostly just leaving holes in various fields.

Although never put into practice, the local volunteers in each parish had to ensure that if animals – possibly even their own – were injured or killed as a result of air raids, they would be humanely despatched.

The Home Guard was a voluntary organisation, but most joined. One NFU member who had a dairy farm was exempted from attending Home Guard parades due to the long hours he had to work on his farm.

For those on duty, the plan was that village church bells

were rung when an invasion was threatened. This would activate the Home Guard. However, the frailty of the system was exposed on one occasion when there was a false alarm in Cameron parish. In the event, the wind was in the wrong direction and few heard the warning bells.

In country areas, the Home Guard was given responsibility for ensuring the blackout was observed. German planes could follow lights on the ground so there was a Government law setting out the hours of the blackout. This was published in the local press, as were the names of those who had broken the restrictions, including people living on farms.

A request was then made in the local paper that all traditionally whitewashed houses and farm buildings such as byres should be painted in a less obvious colour. This suggestion, however, does not seem to have been followed up.

One of the serious aspects of the blackout restrictions was that, in wartime, farmers were unable to put their storm or hurricane lights out in the lambing pens where they could ward off foxes during the lambing season. To get round the blackout regulations, the operators of some dairies found out that 'blue light', or light from only one part of the spectrum, could not be seen by aircraft.

Those defending the East Coast shores also played their part in ensuring the dreaded Hun did not gain a foothold on this island. After days working in the fields, at seeding, at harvest, in winter and in summer, the farmers and their workforce would then be on duty.

The Home Guard believed they were playing their part in the war effort. Physical evidence of their efforts could be seen throughout the countryside. On flat fields, telegraph poles were erected to prevent gliders coming in to land. Farmers were advised to give up their habits of building stooks and hay ricks in neat and tidy rows and scatter them about the fields, all to prevent enemy gliders coming in.

A similar operation to deter invading troops saw trenches being dug on some flat, boggy land up in the riggin, or high parts, of Fife. As these trenches filled up with water from the surrounding land, this latter action had the un-intended effect of being attractive to ducks. The Home Guard had some real shooting practice during the duck-shooting season.

If there had been an invasion, the German army might, in the early days at least, have been somewhat surprised by the Home Guard's lack of manpower and artillery. One unit, based at Cairngreen Radio Station on the outskirts of Cupar, had three men on duty per shift and between them they had only one rifle and five rounds of ammunition.

During training, the Home Guard, consisting of plough-men, butchers, bakers and every other non-combative occupation, had to use wooden rifles to perfect their parade ground techniques. But they also practised with live grenades and there was at least one occasion when the excitement of handling a live grenade was too much for the non-military. Only the swift action of the sergeant throwing the grenade away prevented more damage than the Germans ever did.

Agricultural Executive Committees

In order to galvanise the agricultural industry and increase the production of food, the Government set up Agricul-tural Executive Committees. To underline the importance of the AECs, each of them was allocated a typewriter, a duplicator and the services of one department secretary to operate them.

Just as in World War I, a hefty reliance on imported food was shown to have its weaknesses as German

submarines destroyed many of the merchant vessels bringing supplies from the Commonwealth and the Americas.
Vigorous campaigns aimed at increased self-sufficiency in
food were launched, the primary aim being to increase
food production on the island fortress that was Britain in
the war years.

My grandfather, John Arbuckle, Luthrie, chaired many
of the AEC meetings and the decisions he and those on the
committee took could have far-reaching implications.
Farmers were told to plough up fields, especially for crops
such as sugar beet and potatoes. The acreage of potatoes
grown in Britain in the war years almost doubled when
comparison is made between the early 1930s and the mid
1940s. The scale of the increase in production can be
gauged by the figures from North-East Fife, a relatively
small part of Scotland and an even smaller part of the UK.

In 1939, the area under crop was 68,824 acres. This rose
to 93,362 acres by 1942. On some farms a 100% increase in
acreage was achieved.

Such was the concentration on looking at every aspect
of crop production that a project was carried out in the
Cupar area in 1941 to see if potato haulm could be used as
a base material for making paper. However, despite the
dried haulm having a fibre content of between 25 and
28%, this initiative was never carried through.

The emphasis on the cropping side of agriculture was
further underlined when the Government sent a message
pointing out that 100 acres (40 hectares) of potatoes could
feed more than 400 people. That same acreage under grass
and feeding beef cattle would feed one tenth of the figure.

Coupons for animal rations were issued in 1941. These
were based on one-third of previous usage and brought the
inevitable outcry, especially from those with pigs and
poultry, who required a great deal of bought-in animal

feed. There was also a problem of farmers growing grain being allowed to keep only one-third of the crop. This, as they pointed out, did not provide the necessary protein needed in animal rations. However, the reality was that there was no way the UK was able to import the necessary animal food. While such imports amounted to 1,903,000 tons, or 1,933,537 tonnes, in 1939, by 1943 this had dropped to 11,943 tons, or 12,134 tonnes, of fishmeal from Iceland.

Many local golf courses, including Rossie, Auchtermuchty and Woodmill, Newburgh went under the plough and were never seen again, although, some sixty years later, it is still possible to see the greens and the tees. Even the hallowed golf courses at St Andrews were not immune to helping the war effort, as in 1941 the AEC agreed that they should be opened up for sheep grazing.

For the first time ever, mechanical power was used extensively to increase the area of cultivated land. The Gyrotiller was created, with rotating blades that churned up the soil. This broke up, for the first time ever, thousands of acres of long-term grassland, but as the years went on, the use of the Gyrotiller fell away as it was accused of bringing up sub-soil.

Any farmer resisting the subsidy of £2 per acre for ploughing up grassland was instructed by the AEC that the acreage had to be ploughed. The end result was that more land than ever before was under cultivation.

One of the early landowners to fall foul of the AEC was a Mr Hugh Duncan of Letham Farm who admitted that he did not have the resources, physical or financial, to carry out the work. The AEC undertook the work on his behalf. Shortly afterwards, the same committee was called to the farm of Easter Upper Urquhart at Gateside, where an inspection found it in a 'deplorable state being grossly mismanaged.' The previous year's potato crop was still in the ground and some extremely thin livestock with no food were described as 'merely existing'.

The AEC recommended the Secretary of State requisition the farm. This instruction was never carried out, but one AEC committee member took over the management of the property. Similar action, which needless to say was extremely unpopular with the victim, was taken at several other units as the war progressed. Only 73 holdings out of the 70,000 in Scotland at the time were taken over by the AEC at the end of the war, though.

Wartime workforce

The appointees on the committee, all landowners or farmers, also made decisions as to whether young men who had been called up to defend the country were required to stay at home and 'dig for victory'.

By the early months of 1940, the agenda for the Fife AEC meeting shows a lengthy list of those appealing against going into the uniforms of the fighting forces. Like coal mining and steel making, agriculture was a reserved occupation and those working in it could avoid going off to war. The appeal always came from the employer and a variety of reasons were put forward. It could be that a 'spare pair of horses' was on the farm and the farmer unable to engage any other labour to carry out the work.

In fact, a meeting of the Cupar NFU branch in 1939 asked that all top horsemen should be considered to be in a reserved occupation. Some months later, in January 1940, the Anstruther branch of the NFU reported seventeen pairs of horses were lying idle on farms because the workers had been called up for military service.

Some farmers referred to the pull of employment from the construction industry creating Crail aerodrome as being responsible for shortage of workers on local farms. Others stated a willingness to plough up more ground, if only labour could be spared to carry out the task.

All agricultural workers were subject to the Standstill Order, which prevented employees from moving from farm to farm. This effectively killed off the feeing markets. The Order remained in place throughout the war and was only repealed long after hostilities ceased.

Appeals to avoid war service also came from outwith direct farming, with noted bee appliance suppliers, Steele & Brodie of Wormit, asking for exemption for a worker skilled in the making of beehives. A similar request was made for a wheelwright whose occupation was making and repairing the metal rimmed, wooden wheels of horse carts. Thus, the committee decided on those who stayed behind to produce more food and those considered non-essential to the farms on which they worked.

In some cases, these decisions were a matter of life and death – a dramatic statement, but one borne out by the fact that if someone did not come back from war duty following a decision by the AEC, then it was clear who was to blame.

In cases where farms suffered a labour shortage, the Women's Land Army could be brought in. The WLA came into being in the latter years of World War I. Initially there was a moral concern over sending girls out to farms and cottages in the countryside. This resulted in local branches of the WLA setting out rules, which among other things barred members from entering public houses and smoking on duty. At first, there was also an issue about the availability of the uniform they were expected to wear. The Minister for Defence was asked in Parliament about free clothing for the WLA. His reply stated that more of the existing stores could have been used, but the 'configuration' of WLA members was different from those of other servicemen.

However, the Government must have appreciated their efforts in World War I because the WLA was mobilised in 1939, even before war was declared. Before the conflict was over, more than 200,000 volunteers had added farm work to their knowledge and skills. In their khaki tunics and wide-brimmed hats, they were a familiar sight in the countryside. Although often left with the more menial tasks, they also drove tractors and worked with horses. Feeding hens and livestock was a more common occupation for WLA personnel.

The Cupar branch of the NFU wanted to see the Women's Land Army organised on military lines so that squads of WLA could be sent out in batches at relatively short notice.

The WLA were either billeted on the farm on which they worked, or housed in a number of large houses around the countryside. There were matrons appointed in the hostels, as well as a senior girl with deputies. These deputies often drove the girls to work and trained them in farmwork. Some of the WLA also became skilled in ratcatching – these pests ate and spoiled many tons of desperately needed wheat and potatoes. Their efforts in home food production continued well after the end of the war and the WLA was only disbanded in 1950.

While the WLA helped reduce the shortage of labour on farms during the war period, there were still major problems, particularly at harvest time and potato lifting. Early in the war, students at St Andrews University applied to work at the harvest but this help dried up as more and more undergraduates were called up. In the depths of 1941, with so many men called away, the NFU appealed to the Council to release roadmen to help bring home the harvest. The Council's response was that the numbers of roadmen were already depleted through service call-ups and maintenance of the road network was equally important to the war effort.

Also in 1941, as part of the war effort, the summer school holidays were cut to four weeks, with the schools going back in August and then closing for the month of October to help with the potato harvest. The rector of Bell Baxter School in Cupar described this move as very detrimental to education: 'We cannot educate them and let them off for potato gathering as well. In the lower classes, the incoming material – assumed to be pupils from our feeder schools – is lower than it has been,' he fumed.

The move to a shorter summer school holiday in 1941 affected the raspberry harvest, with fruit growers objecting to the loss of their workers in the traditional picking period. As a result, some thirty pupils at Bell Baxter were granted exemption.

During the early years of World War II, Fife Education Committee did not strive to enforce the law that prevented the employment of youngsters under 12 years old. However, as one member of the committee remarked, 'If it comes to 5-year-olds working, then action would be taken.'

Help on the work front then came from another source: prisoners of war. Initially they were guarded and a young John Purvis, now a member of the European Parliament, recalls seeing armed guards looking after a squad of POWs picking potatoes. By 1942, good conduct prisoners were allowed to stay on farms and were paid between 6d and 1/-, or 2.5 to 5 pence, per day.

Mostly, though, they were transported daily from the three camps in North-East Fife at Annsmuir, Ladybank, at Bonnyton and Lathocker, both outside St Andrews.

Soon there were 37,000 POWs on farms throughout the UK, but the labour shortage was still acute, and in 1943, the Ministry of Food asked for a further 36,000 to be released from their camps so they could help produce more food.

At first, the majority of POWs were Italian, but by 1944, German POWs were also down on the farm, with the farmer keeping a watch on the two nationalities, with their very different temperaments.

In some cases, hardcore Nazis refused to work and a group went on strike in 1947. This action followed complaints about the quality of labour provided from the camps.

In one Union minute, it was reported that a 'lower grade of men of the Nazi type' now had to be employed. Cupar NFU, in 1945, discussed whether they should have to take German POWs without armed guards. The decision was that they could take one or two.

Farmers had differing opinions of their enforced labour, with many seeing the Italians as lazy and only interested in singing all day. One farmer in Fife considered this, his introduction to Italian opera, a classic example of the law of unintended consequences.

By 1944, when the worst of the labour shortage was over, the Cupar branch of the NFU heard complaints that amenities in the prisoner-of-war camps were better than those on the farms. It was said that it was not surprising Italian prisoners preferred to stay in the camp, rather than come out to work. The allegations continued, with statements that the prisoners were also allowed to go to the cinema twice a week. Another complaint was that the prisoners now had to be taken to doctors, dentists and hairdressers, and have their clothes seen to. The disgruntled Union member reckoned that farmers were merely acting as flunkies to the former enemy. He was further put out by the fact that the Italians did not appreciate British food as they preferred their own spaghetti and pizza.

In another example of a change of attitude, in 1946 one of the NFU members complained that where POWs were working in sugar-beet thinning gangs, they were only doing approximately one-fifth of the work done by the farm staff. To back it up, he gave figures that showed the POWs had singled two drills each, which meant that they did 500 yards a day compared with the 2,600 yards achieved by the farm staff. 'Even then, the quality of work was poor,' he added.

On the other hand, there were also farmers concerned over the quality and quantity of food the POWs were receiving and they were keen to provide them with hot soup or tea during the day. But it was pointed out that giving food to prisoners-of-war was an offence which would result in the withdrawal of the prisoners from the offending employer's service. After this decision all food was supplied by the camp.

As the war ended, repatriation commenced and again the change brought problems for the farmers. During mid-harvest, September 1946, it was suddenly announced that the 700 German prisoners at Dunino prisoner-of-war camp were to be repatriated. The fact that their place would then be taken by 1,500 Poles did not satisfy the farmers in mid-harvest and they decided to protest about a move that would cause about a week's lost labour. There was also an issue at Annsmuir, where in 1945 there were 750 German POWs quartered. Farmers hoped the authorities would increase this number, but they were unable to find the accommodation. Therefore, as far as farmwork was concerned, farmers were advised that they should make use of Polish soldiers stationed in the area.

Later in the war and also in the post-war period, the

farm workforce was further augmented by Displaced Persons, or DPs; this unfortunate group of people had either seen their homelands engulfed in a new political regime or had decided mainland Europe was so war-torn and poor, they would be better off living and working in this country. DPs mostly came from Eastern European states on the far side of the newly created Iron Curtain.

Other workers came along in the aftermath of war. They were called European Voluntary Workers, but farmers complained there was a 'slackness and indiscipline in this workforce'. Many of these DPs and EVWs stayed and continued to work on farms until the early 1950s. Unlike the next wave of European workers who came over in the 1990s, they had to work in agriculture.

The Annsmuir camp closed its doors in 1952 and is now a permanent caravan park.

The control of labour was only a small part of the work of the AEC committee, who were also given the task of putting existing machinery to best use. The Government supplied extra harvesting equipment under the grandly titled, Government Tractor Service, and it was left to the AEC to determine how this was deployed. Locally, they decided maximum output could be achieved if teams of five workers accompanied each of the Government-supplied binders. The team consisted of one female tractor driver, one male binder man and three Women's Land Army members for stooking. However, this was too much for one farmer, who complained to the Union that the Government was paying these teams higher rates than normal farm workers.

Subsequently, following the late arrival of the USA into the war, a deal was struck in 1941 between President Roosevelt and Winston Churchill that resulted in thousands of tractors and hundreds of combines, binders and balers being brought over from the States, with payment deferred until after hostilities ceased. Churchill called the deal, 'the most un-sordid act in history.'

At the beginning of the war there were 55,000 tractors in the UK, six years later that figure quadrupled to more than 200,000. Combine harvester usage went from 1,000 at the beginning of the war to three times that figure by the end.

To the current day, there are vintage tractors that came over on the 'lend lease' deal that helped this country survive. Makes such as John Deere, Allis-Chalmers, Case, Minneapolis Moline, Caterpillar and McCormick International made their first arrival as part of shipments from the USA. As it was left to the AEC to determine just who would receive these tractors, some decisions set neighbour against neighbour.

After the conflict
While there were universal celebrations at the cessation of hostilities and many a Victory Day party or barn dance was held, the legacy of the war years continued to dominate the rural scene for most of the next decade. There were shortages of almost every commodity, as the wheels of industry had been turned to feeding the hungry maw of war.

Farmers applying for extra rations of petrol to go to the 1948 Highland Show in Inverness had their request refused as being for non-essential travel. In order to justify a visit to the local town, they would often put a bale of hay or even a calf in a sack in the boot of their car. If it was a Saturday night, those living in the countryside would leave their vehicles at a farm just on the outskirts of the town and then walk the last mile or two rather than contravene the rules relating to fuel rations.

In 1948 it was stated at a Union meeting that there was no harm in disclosing to Petroleum Officers that petrol supplies for tractors and other farm machinery had also been used for cars in connection with the farm. One farmer who did not heed the rules ended up in the local Sheriff Court, where he met a very unsympathetic Sheriff, who complained, 'I get a miserable pittance of a fuel allowance and I have to get lifts from farmers when I have no petrol.' Unsurprisingly, the verdict was not in favour of the accused.

The Union debated the shortage of sodium chlorate required for the burning-off, or stopping, the potato crop growing. In this case, the Board of Trade made arrangements to send up sufficient chlorate.

Lengthy debates were then held at NFUS meetings over the shortage of Thermos flasks. For those who always have a kettle to hand this might seem somewhat frivolous, but if you were out on a cold and frosty morning on a tractor without a cab, or were pulling turnips with a frosty hoar on the haulms, you welcomed the hot drink the Thermos provided. Farmers had had priority ordering for Thermos flasks during the war but there were problems when questions were asked whether potato workers should be entitled to a Thermos flask permit. The supply of Thermos flasks then became one of the headline issues the Union took up with the government of the day.

Even three years after hostilities ceased, there was still a shortage, with an NFUS meeting hearing in 1948 that there were some 250 flasks on order but that it could take up to six years to get delivery. However, one of the more travelled members blew that away by saying Thermos flasks could readily be bought in London for 7/6d, or 37.5 pence.

Even in the early 1950s, shortages of various materials

were reported. In 1951 an extreme shortage of baling wire was noted. Coal was scarce and in 1950 dairy farmers appealed for extra coal to permit the sterilisation of their milking equipment. The milkmen were subsequently supplied with a lower grade of coal, unsuitable for domestic purposes.

Learning from their experience in World War I, the Government quickly introduced a comprehensive food policy right at the start of hostilities. Food ration books were issued and the basic rationing of such commodities as eggs, cheese and meat, along with more frivolous items such as chocolate continued right up until 1952. Food coupons allowed everyone an equal chance of buying food – including eggs and meat.

Those farmers with milk, eggs and poultry were often suspected of selling some of their produce direct to customers, thus bypassing the official Government-buys-all-food policy. A decade after hostilities ceased, farmers with permits to sell eggs to the public had their records checked to ensure no black market trading was being carried out. The local paper reported many incidences of 'Black Marketing' of food and in some of these cases, farmers ended up in the dock.

When rationing started, farm workers joined the exalted company of charcoal burners in being allowed an extra ration of cheese – 8 ounces, or 200 grammes – every week. However, this made the farmers unhappy and the NFUS petitioned for them to receive the additional ration as well.

The food rationing policy may have been unpopular, but it was successful. At the start of the war, the UK imported 60% of its food, but by the end, despite imports being halved, the nation was still fed.

This ability to feed the nation from within its shores

provided the most significant piece of agricultural legislation in the last century: the 1947 Agriculture Act. The
Government of the day, under the leadership of Clement
Attlee, brought in policies that would provide subsidies in
the form of guaranteed prices and deficiency payments for
the farming industry. Attlee's aim was to further increase
home production by another 20%, more than half as much
again compared with pre-war home supplies.

This introduction of widespread financial support for
farming helped convert the industry in this country into
one of the most efficient and productive food suppliers in
the world.

Before we leave war behind, it is interesting to note that
in 1959 Cupar NFU was treated to a talk on the
after-effects of atomic warfare and how damage could be
minimised on the farm from any nuclear holocaust.
Thankfully, the recommendations given at the meeting
have never been tested.

Chapter 23

Transport

IT may have seemed mundane to my schoolboy colleagues who lived in towns, where the delights of picture houses were always on their doorsteps, but the Saturday bus that meandered along the rural roads of North-East Fife before disgorging its passenger in Cupar was, for me, an opening to another world.

Right up until the private motorcar became universally available in the 1960s and '70s, the rural bus services brought the country to the town. Thus my travelling companions were the workers' wives going into town for the shopping. Travelling vans were still common and they brought most of the requirements to the farms and cottages but there was a realisation that small commercial vans did not bring the range of goodies available in towns.

Also travelling, were the menfolk who decided that a visit to one of the public houses in town would slake a week's thirst and also provide the best gossip. Among the latter group were the Irishmen, who were in the area to pull sugar beet. On Saturdays they would go into town to shop and possibly go into the local hostelry. Then, on Sunday mornings, they would get the first bus into Cupar to attend chapel.

In the pre- and immediate post-war days, many of the buses still had wooden slatted seats, and as the bus swung round the corners of the country roads, the passengers slid to and fro along the seats.

Without a great deal of cause or reason, youngsters travelling on these buses were invariably in awe of the bus conductress. They had seen how she could gather her passengers out of the pub to get the bus away on time and they had also witnessed her dismissal from the bus of some unfortunate tramp who did not have the fare. But there was also a kindlier side to those uniformed guardians of the buses. Regularly, on the late-night Saturday bus – by this time it was after 10 p.m. and that was late – the bus would stop at a single estate cottage deep in the country. The driver would come out of his little cab and gently take his father, who had been too long in the pub, off the bus. Along with the conductress, they would help the elderly man across the road and into his cottage. At this point the conductress returned to the bus, where all the passengers sat quietly. There was no 'bus rage' about the involuntary stop. A good few minutes later, the driver clambered back into his cab, having ensured his father was tucked up for the night.

At the same time as I was undertaking my youthful journeys, little buses would be traversing most of the country roads in Fife. But, whatever level of service was provided it never seemed to answer all the requests. Many a County Council and National Farmers Union meeting had an item on bus services and these invariably produced discussion and dispute but never resolution.

The heyday of the rural bus service was from the 1930s until the early 1970s. In the early days, it opened up a wider world for those living in the countryside.

Walking and biking
Prior to this public service becoming available, the working family had little or no opportunity to venture further afield, even to the local towns.

They were largely dependent on Shanks's pony, a term that has slipped away as fewer and fewer people walk and therefore understand that Shanks's pony meant that you simply walked. And many recollections include walks that seem incredible to today's more sedentary population. People would walk from deep down in the East Neuk to St Andrews. Those living within ten miles from Cupar would see nothing unusual in a walk into the market town.

A step up from walking came with the bicycle. These were not new, but few could afford them in the early days. Often a worker's bicycle would be a treasured possession. The bikes they used were strong and sturdy; there were no gears to aid the travel other than varying speeds in the legs of the owner.

Bikes would be used for leisure. A Sunday afternoon tour of the neighbourhood would allow the farm worker to check up on the progress or otherwise on farms in the vicinity. Such trips formed the basis of gossip for the rest of the week.

Bikes would also be used for carrying goods. Visits to local village shops would inevitably mean a laden bike being ridden precariously on the return journey.

Horse and car traffic

The only traffic on the sparsely used roads in those early years were either wagons or commercial vans carting goods to and from farms or some of the upper classes out with their Broughams, carriages or gigs.

Commercial trade was far more limited than it is today. Needs were much fewer and most farms were very self-sufficient and self-contained. Their ingoings would include a few tons of fertiliser brought by boat to one of the coastal ports, or by railway wagon to the nearest station. There, it would be met by local contractors or even the farm's own carts for the last leg of a long journey. If the journey was relatively short, less than five miles, then the farm worker with a single horse in a coup cart could take about a ton of potatoes, grain or fertiliser. When the journey was longer, or if more tonnage was to be transported, then a team of four horses would be harnessed into a four-wheel wagon capable of carrying up to 5 tons.

In passing, think of the use of fertiliser brought in bulk to Dundee or Methil. There, it would be hand-loaded into railway wagons and then hand emptied into the contractors' or farmers' carts. That is one reason why there were vast numbers employed in the countryside a century ago. Seed grain and potatoes would then make the return journey by horse cart, by rail, and often seed potatoes, would go by boat to the potato-growing east coast of England. But overall, the tonnage of produce heading along the country roads was only a fraction of today's traffic.

There would also be vans supplying groceries, bakery goods, butcher's meat, fish, oatmeal, milk, fruit and vegetables pulled from farm to farm by a single horse. These would be weekly rounds undertaken by local shopkeepers, as an addition to their shop trade. Their range would be limited by the speed of the horse and as this was reckoned to be no more than an average of two miles per hour, the number of customers on each round was quite small. As they travelled their slow road around the area, these commercial travellers would be overtaken by the light-legged, high-stepping horses and ponies used to pull the gigs and carriages.

In the gigs would be the lords and ladies of the day, or the professional classes – the doctors, bankers and lawyers going out to see their customers, or at least all those who could afford to pay for their advice.

Farmers often rode their own steeds into market, with livestock being sent off earlier in the day under the watchful gaze of the shepherd or cattleman. He would have the doubtful advantage of help from the farm loons, who used to stand in open gateways to prevent the livestock heading into fields full of crops. There would be an even bigger pantomime in the droving when going through small villages.

Even if the days of long-distance droving were long gone, livestock travelled miles and miles. Trips of eight and ten miles to market were not uncommon and, for the stockmen, there was the added concern that the farmer might also buy another lot or two for the return journey.

These sheep and cattle could come in any condition. If they were Irish, or even Canadian store cattle, they might well have been in transit for a couple of weeks; if they had been well looked after, they would let loose and the first mile or so saw the herdsmen in hot pursuit. However, if they had been treated badly, often they were in a weakened condition and it was difficult to persuade them to keep moving. A tendency to sit down in the middle of the road posed problems for herdsmen intent on getting back to base before sunset, as well as for passing motorists. To add insult to injury, it always seemed the above scenario occurred just as the boss swept into sight on his journey homeward from the market and then the public house, where the purchase had been sealed with drink taken.

In 1938 the *Scottish Farmer* noted one drawback to the move to motorised transport: 'With all respect to the memories of the older generation, sales and market are often now more sober affairs. Homeward journeys have to be done at the wheel of a car while Dobbin could, with little guidance, find his way back to his own manger.' It was to be a number of years before strict laws on drink driving were introduced, but the warning came in with the first of the cars.

Before the large-scale adoption of the motorcar and lorries, livestock were also sometimes transported by floats, pulled either by oxen or by teams of horses. This was not a common practice and was often restricted to moving valuable bulls or stallions around the countryside.

One of the busiest times of travel on the roads occurred at term time when men, their families and all their worldly possessions would be heaped onto a cart as they moved to their new home. Farmers also tended to move base more often than is the case nowadays. For those changing farm, pre-mechanisation, there was surprisingly little to move.

Some moved farm with more style than others: in 1922, George Watt moved from his family farm at Wester Kilmany to a new base at Cornhill, Collessie, a distance of some eight miles. The journey was made easier for him, and more pleasant for his horse, and for those he passed on the way by his playing of a piano while travelling. History does not record the tunes he belted out on his flitting, but suggestions include, 'Show Me the Way to Go Home' and 'It's a Long Way to Tipperary'.

For many farmers, the arrival of the car widened the horizons and early open vehicles soon replaced horses and gigs. Traffic was sparse in those days and there were few problems of congestion in the area. However, one of the first parts of Cupar to restrict parking was around the Auction Mart. In 1948, the congestion caused on market days with livestock vehicles coming and going had to be controlled. In a rare show of humanity from a local authority, Cupar Town Council made an exemption from the no-parking edict around the Cupar mart for Mr Thom, of Bunzion, Ladybank. In making the case, it was stated that Mr Thom had attended the market for the previous sixty-five years and he was no longer able to walk very well.

In an era where the phenomenon of the white-van man rushing about the country is widespread, it is curious to recall that only fifty years ago, there was a law that all pick-up trucks were limited to a maximum speed of 30 mph.

Road conditions

A report in the latter half of the nineteenth century notes that Fife was particularly well supplied with roads. This applied not only to the turnpike roads where the income was generally in line with the expenditure required.

One possible reason for the good quality of the road network in Fife might lie in the ready abundance of stone and quarries in the county. For well over a century, some of the hard stone or whinstone quarries have been worked. In that time, millions of tons of material has not only gone onto the road network in Fife but has also been shipped in coasters to ports in the south of England and as far as Europe.

During the early years of using tar to hold the stones together, as invented by John Loudon MacAdam (1756–1836), there were a few teething problems. In 1922, the Anstruther branch of the National Farmers Union wrote to the Council regarding the slippery and dangerous state of road surfaces. The Union's ire was raised because most of the country roads were laid out with two hard ridges of track, along which the wheels of the carts, gigs and carriages would travel. The centre of the road, where the pulling horses would be was now covered in tar and chips; these were being spread across the entire road with no track left in the centre for horse traffic.

Following the virtual demolition of the road between

Alloa and Dunfermline in 1924, the County Council proposed banning heavy vehicles from the roads in the summer months unless they were moved before the sun came up and melted the tar. The National Farmers Union was asked to support the argument put forward by the Scottish Traction Engine Owners and Users Association that this would be a restriction on trade. The fear was that farms would be at a standstill if travelling mills driven by large steam-powered engines did not have free passage at any hour of day.

After it was pointed out that there was precious little harvesting in the summer months and that most of the mills travelled late at night or early in the morning, the Union decided not to oppose the byelaws. However, the use of tractors and implements on the public highways has been a constant source of concern throughout the past century.

In the early days of tractors, most were kitted out with steel wheels. Often these had diagonal bars welded on them to prevent slippage in the fields. No one had calculated just how much damage these same steel wheels, with their ribs of iron, could cause to roads.

A more extreme method of preventing slip in the plough furrow on the steep incline came with spud lugs, which were lumps of metal regularly placed around the perimeter of the wheel. Again, the indentations in the field might have been beneficial, but shifting these vehicles across the carriageways from field to field either involved illegality or the use of battens that would be laid below the wheels.

The fact that the Council was not amused by the damage to its road network can be gauged by the prosecution of a farmer in 1938 for allowing his tractor, without pneumatic tyres, to travel along the public highway.

As a youngster heading downwards to school, I can recall the ridging of Caterpillar tracks scoring across the metalled road.

One consequence of bad weather at harvest time also brought a source of irritation between farmers and members of the public that continues to this day. This is the issue of mud on the road. It was virtually impossible to move from sugar beet fields enveloped in deep glaur to the farm steading without leaving a fairly obvious trail of mud. In addition, the weekend might have seen the family car scrupulously polished until it gleamed; so it remained, or at least until it careered into the aforementioned mud.

Whenever there was a problem at a particular location, the roadman would make his appearance. Known as 'section men', they looked after the verge cutting, drain cleaning and general maintenance of their section of road. The local section man in my schoolboy days was Willie. We would see him arrive with the spades or shovels he required for the day's work slung about his bike.

More dangerously, on days when grass needed cutting, he wrapped the scythe around himself and his bike, and proceeded to work. All this was before the existence of the Health and Safety Executive, otherwise they would most likely have stopped the practice. I am happy to report no accident took place and Willie retired, all limbs intact. However, it was noticeable that when he was cutting the verges, he stashed his scythe away at the end of a day's work rather than increase the odds of damage and disfigurement.

Ferries

The geography of Fife may be considered a hindrance to its economic development. Prior to the arrival of both the Forth Rail Bridge and its Northern counterpart, the Tay

Rail Bridge, in the late nineteenth century, access – particularly to North Fife – was limited to ferry crossings.

Although they no longer had a monopoly, the ferries continued to ply their trade across both Forth and Tay for the better part of the twentieth century. And it was from an incident on one of those ferry trips that the nascent National Farmers Union cut its teeth. Paddy Henderson, Vicarsford, had been transporting ten homebred cattle on the ferry when they took to the water as the steamer approached Dundee. A lifeboat was launched, and with the help of the three lifeboat men, the cattle were chased onto the shore. For their efforts, the lifeboat men who had rowed their boat some three hours in the rescue bid were given 5 shillings each, Mr Henderson saying they wanted no more as it had been 'good sport'. However, they must have reconsidered their position, and through their employers, the Dundee Harbour Board, they put in a claim for £75 against Mr Henderson.

The Union took up the battle on the basis that the cattle were actually under the control of the ferry owners, the Harbour Board, when they made their escape. The Sheriff took the side of the three men, leaving the Union to face the legal bill.

For those who did take their cattle over to market, the stock would mingle on the deck of the ferry. Only on alighting at Dock Street, Dundee, would the owners split the bunch up prior to taking them into market.

Fife is surrounded by water on three sides, but surprisingly little agricultural trade was conducted through docks in Fife. In earlier years, grain would come and go from the docks in Kirkcaldy with two large-scale maltsters using the facilities to import grain or export malt.

In the previous century some seed potatoes would be loaded onto shallow-keeled boats along the shores of both the Forth and Tay. These coasters, taking their loads down to the potato-growing areas in eastern England, would literally come onto the shore at high water and when the tide went out, horses and carts would come alongside and transfer the potatoes onto the boats.

Seed potatoes were also loaded in St Andrews harbour in the early 1920s, and the local Union asked that the Government consider dredging out other harbours in the East Neuk to help further the trade.

Railways

If any one form of transport could be said to have unlocked the economic potential of farming in Fife, it was the arrival of the railway.

The first lines to stretch through the county came in the mid-nineteenth century with the strategically important north-south line going over both the Tay and Forth rivers. For the first time ever, Fife lost the disadvantages of being a virtual peninsula, with land traffic confined to its Western boundary.

The railway breakthrough brought with it an increase in the population of the county. Valuable coal reserves could be tapped and sent onwards to feed the iron and steel industries fuelling the Industrial Revolution. In a quieter, less obvious fashion, the rail tracks heading through the country also enabled an agricultural revolution to take place. This was underlined in the first decade of the twentieth century by the building of rail routes through rural areas such as the East Neuk line and the North of Fife line, where most of the future traffic would be agriculture based. With the spider's web of a rail network now existing throughout Scotland, livestock and crop could be moved more easily.

The importance of the railway network is shown by

the range of transactions at Drumrack Farm, Anstruther: machinery spares came in from Wallace of Glasgow by train; pigs were brought from Hawthorndean Station in the Borders. Likewise, lambs were brought down from the big sheep sale at Lairg. For some unknown reason, there is an entry for 'empty crates' from Hawick. A collection of beehives, bought in Wormit, came down the line; empty chicken boxes returned by train to Stirling. Much-needed fertiliser, nitrate of soda, came from Leith, no doubt shipped in prior to going on the train. A Clydesdale horse arrived from Errol and was then sent back again two weeks later with no explanation on the accounts.

Not content with that, Mr Watson of Drumrack then bought ashes from the railway company for use around the farmyard, especially the stackyard, as they helped keep the bases of stacks dry.

The Highlands and Islands of Scotland have always bred good stock, but until the arrival of the train there had been no easy or quick method for getting these cattle and sheep onto the lower, easier ground for fattening. The old droving system could not be described as easy or quick.

With the arrival of the train many Fife farmers made annual pilgrimages to the auction marts held in the north and west of Scotland to buy lambs and store cattle. These would then be finished on the turnips and straw that existed in abundance in Fife. Even those farmers who did not travel would often have an agent or the auctioneer to buy on their behalf and then put the livestock on the train.

It was an efficient service: livestock could be transported anywhere in the country. The last leg of the journey for store stock from Ireland into Fife was reckoned to be one of the shortest as far as time was concerned. And there were cases of cattle being delivered to the farm before the farmer himself had returned home.

That was definitely the case when Andrew Peddie of Coal Farm, St Monans bought a bull at the famous Perth bull sales. He arrived back to find that the stationmaster at Anstruther had already been on the phone wondering when he was going to collect his beast.

Trains also played a vital part in the development of other enterprises. They brought wagons that would take the milk to Dundee and Edinburgh on a daily basis. Cans of fresh milk would be loaded at little stations like Mount Melville, Luthrie, St Fort before being carted off to Dundee, Kirkcaldy or even Edinburgh.

The grain trade saw its horizons expand with the rail network and the rail companies also noted the potential for added business with this crop. They bought thousands of sacks to help in the handling of the wheat, oats and barley, which they would hire out to farmers. But it was not a business without concerns; invariably there were insufficient sacks during the milling season, or the sacks were at the wrong location. Complaints flew between farmers and the rail companies over the condition of the sacks, with farmers often complaining about the state of the heavy twill sacks being used. They could not have been too bad since, late in the 1960s, it was still possible to see railway sacks with dates going back into the 1930s stamped on them.

The railway companies also had their moans as they objected to the farmers letting vermin, such as rats and mice, play havoc with their valuable sacks.

Another use of the railway sack was as a hood for ploughmen sent out to work in the rain. Today's generation of 'hoodies' may use the hoods of their jackets to prevent identification, but their predecessors used them to help keep dry.

Later, as road transport took more of the farm produce,

the grain and potato merchants would supply sacks. An ideal system, it might be thought, but one which could quite easily go wrong, with the farmer filling the wrong sacks or sending the produce to another merchant.

The potato trade also received a massive boost with the arrival of the railway. For seed growers, whose market for healthy stock was in the east of England, wagons could be loaded locally and then freighted down to areas that could grow large tonnages. Ware or eating potatoes were also loaded onto railway wagons. In the early days, the vans would be loaded in bulk, and the potatoes graiped into the wagons. But this system soon fell into disrepute. In 1935, a complaint was made at Cupar branch of the National Farmers Union to the effect that there had been a loss of tonnage. In this instance, wagons sent from Luthrie station, which incidentally handled more potatoes than any other on the rail network, had been sent to Bathgate, where, on emptying, they had been 10 hundredweight short. Suspicions fell first on the merchant and then on the rail staff, who it was claimed by the Union member, 'were not devoid of taking such an opportunity.' No conclusion was reached.

Frost possibly played a bigger part in stopping this practice: these bulk vans could not easily be protected against the winter elements. Even when sacks were used for the potato crop, the farmer had to provide the straw for protection against frost damage. When the weather was particularly severe, all rail loadings would cease. Once that occurred, there was always a 'post-freeze' squabble over who had to pay for the wagons while they lay idle. A farmer receiving a bill for this demurrage was never a happy person.

During World War II an edict came out from Government stating that railway wagons had to have only one sheet of cover. When the NFU pointed out that potatoes were easily affected by frost, the response from the Minister of Transport stated he did not see why potatoes should have preferential treatment over anything else.

When the weather was suitable, the farm carts and trailers arrived and farm workers would load potato sacks while the porter would be the 'straw man'. Often there were queues of farm vehicles waiting to be unloaded or loaded at goods sidings and the allocation of wagons was often a point of contention, especially at sugar beet and seed potato movement time.

Frequently, there was also the issue of getting the vans into the sidings. The trains would drop off the empty vans, but it was often left to the horses or tractors to shunt them into the loading bays.

An example of the extent of use farms made of the rural rail network occurs when the old station at Leuchars was closed down for passenger traffic in 1921. All the rail staff, except for 'one oldish porter', moved to the new station. The National Farmers Union complained to the rail company that the old porter would have far too much to do, with some twenty to twenty-five wagon loads every week during the winter period.

It was not only farm produce that benefited from the network. Every night the East Neuk line would take two wagons of lobsters to the London market. During the fruit picking season, the afternoon train would stop at Cupar to collect freshly picked strawberries heading south for the big Smithfield market. This train would also have stopped to pick up fruit at all the main berry picking areas in Angus. However, this domination of the transport scene soon faded for the rail companies. A lack of investment and the increased flexibility given by the motorcar and lorry saw branch lines such as the East Neuk being chopped off by the early 1960s, after a mere fifty years of existence – but one which had transformed the area it served.

A decade earlier, the North Fife line had closed to passenger traffic and muted opposition to this move was soon quietened by figures from the rail company showing only ten passengers using the up-line to Perth, while six used the down-line. The cost to British Railways of running the service was put at £2,707.

There is now no transport of agricultural goods by rail in Fife.

Lorry transport

With development of petrol and diesel engines, it was inevitable that more and more traffic would end up on public roads. By the mid-1950s, it was reckoned that half the sugar beet being delivered to the factory was coming by road. Rail wagons still played an important part, but the move was definitely towards heavy goods vehicles.

Initially, a 5-ton lorry was considered to be top-of-the-range and the line-up waiting to go into the factory saw many such lorries waiting to be emptied alongside farm trailers, with their 3-ton of beet. With improvements in the road network, lorries could, by the 1960s, travel down to the south of England much more efficiently and quickly than via a rail network starved of investment for too many years.

The arrival of double-level livestock floats changed the equation, especially as they would deliver from farm to farm, thus eliminating the 'herding extras' previously required in moving stock. For Fife, this move to the road away from rail could have had an adverse effect on the farm economy. In road terms, and as far as north and south traffic was concerned, the county was back to being a peninsula.

This was recognised as far back as 1953 when the Anstruther branch of the NFU recommended support for the commencement of a Forth road bridge. And as a fallback, the NFU wanted British Railways to be pressed to improve the present inadequate service from Queensferry. The local MP, James Henderson Stewart, was unimpressed by the request and did not think a road bridge over the Forth would be possible. He urged that the Burntisland to Grantown ferry be restarted to improve transport links and reduce congestion.

The prescience of the Anstruther Union members was accurate, as those who survived the next decade were able to drive themselves over the Forth; either to Edinburgh or as a gateway to the south.

A similar look into the future took place in 1961, with Cupar branch of the NFU supporting a railway halt at Edinburgh Turnhouse airport. Some fifty years further on, this has still not been achieved. What was gained was a road bridge over the Tay. Its opening, in August 1966, was accompanied by the Scottish College advisor reckoning that this would be a tremendous boon to strawberry and raspberry growing in Fife, as it would have access to fruit processing factories in Angus. Sadly the prediction did not reckon on these factories closing over the next two decades as demand for canned fruit and jam diminished.

However, the road bridges to the north and south of Fife nowadays provide an easy link to the major grain malting premises in Angus and the Lothians. Both provide the main routes into the county for the fertilisers that are now imported, either at Dundee or Leith. Of course cost was a factor in all this. Back in 1937, one member of Cupar NFU reckoned he could transport his requirement of fertiliser around by Perth more cheaply than if he moved it by rail over the Tay bridge.

The last note on transport in Fife might come with a reference to the cunning plan devised by the locals in case they were invaded by the Germans: they took down all the road signs to confuse potential invaders. Half a century on, one wag has commented on the fact that there are still some areas where directional signs have not yet been replaced: 'It's all right – Fifers always know where they are.'

Chapter 24

Trade

THE Corn Exchange in Cupar is an imposing building and, like many other similar properties the length and breadth of the country, its origins lie in the prosperous grain trade of the 1800s. Today, however, its role in trading grain has gone completely and the building is now used for a succession of coffee mornings, model railway exhibitions and musical evenings.

The demise of the Corn Exchange was hastened by the arrival of the combine harvester. In binder and straw stack days, the sale of grain took place over the winter months as part of a more leisurely process. Nowadays, with the speed and capacity of the combines, and the fact that two-thirds of the annual crop is traded in the six-week harvest period there is just no time for a weekly grain market.

It was still operating as a Corn Exchange in the 1960s, when, as part of some work-experience project, I accompanied a grain merchant up the stone steps and into the main hall. Around the perimeter were booths, each with the merchant's name above. The floor was scattered with grain as samples were taken out and inspected, talked up, talked down and eventually traded, and the final deal was accompanied by a handshake: for some of the older merchants, this was preceded by spitting on their hands.

These same merchants regularly supplied seed grain, fertiliser and spray chemicals. This system often led to just an annual settling of accounts with all those costs being deducted by the merchant before he signed a cheque for the grain. Among those operating on this year-round trade, the axiom was that if you had a good grain trade relationship, then the rest would follow: fertilisers, sprays and feedstuffs. In lean years, however, quite a number of farmers were unable to settle their fertiliser accounts until the grain was sold.

On my first appearance in the Corn Exchange, I was somewhat taken aback by a farmer handing me his cheque-book and his bill. My lowly role was to write the cheque for the required amount and pass it back to the farmer for his signature.

In addition to the trade carried out at the Corn Exchange, in the 1920s and 1930s, firms supplying fertilisers or buying potatoes or grain set up networks of salesmen using the newfound flexibility provided by the motorcar. These 'travellers' would find their way up and down farm roads and were generally reckoned to have an infinite capacity to drink tea and sit around the kitchen table, passing on the gossip of the neighbourhood. The best salesman also knew the best gossip.

Such were the numbers employed as travelling salesmen or grain buyers that right up until the 1980s hardly a day passed on a farm without the arrival of at least one visitor. The local co-operative, Central Farmers, had four travellers in Fife alone, each with his set area. In the early 1950s, grain and seed merchanting company, Carnegie and Grant, had up to fifteen representatives.

Unlike today, when most farmers do their business through a single company or co-operative, and by phone or email, their predecessors would have to see a range of potential buyers or sales staff. It may not have been congestion, but often there was a queue of travellers waiting to see the farmer. As a result, trade was frequently split up between firms, and traders would go down the farm road happy with the promise of a load of potatoes or an order for grain seed. At harvest time, the pace would be frantic for travellers who went from farm to farm, collecting grain samples. They would fill special envelopes with grain from each harvested field, then hurry on down to the potential buyers.

Before chemical analysis became available, merchants bought grain by sight. If a sample was rejected, the saying was that, 'It would be better the next day'. Many a lot previously turned away was later traded without issue. At harvest time, offices of grain merchants would almost burst at the seams under the weight of samples waiting to go to the maltsters or feed merchants.

In the first half of the century there were still a number of local maltings, such as Ladybank and Newton of Falkland owned by the Bonthrone family, where the grain would be turned into malt, ready for the distilling market. Now, these have all gone, although the premises at Pitlessie have been converted into housing. Today's residents may not be aware that the sacks of grain were hoisted up by a pulley and chain before the contents were tipped out onto the wooden malting floors.

In Kirkcaldy, where they had the added advantage of access to shipping, there were two large grain-handling premises. In the mid-1960s, R. Hutchison & Co. of Kirkcaldy employed some 200 people in the milling and malting trade. Kirkcaldy was also the base for R. Kilgour & Co., who made malt for Ballantynes whisky. These premises have gone, leaving just one massive distilling operation at Windygates, operated by international drinks company Diageo.

Before leaving the grain trade, it should be noted that recent concerns over the use of foreign grain in the making of Scotch whisky have long been a bone of contention with Scottish cereal growers. As far back as 1938, the East Fife area of the NFU put forward a proposal requiring distillers to make their mash with 100% homegrown barley until home supplies were exhausted. That move, and all subsequent attempts to link the base material in whisky with the name Scotch, have been strongly resisted by the all-powerful Scottish whisky industry.

Part of the farmers' case has always been that they receive only a small fraction of the final sale price of the alcohol. Back in 1930, it was estimated that a bushel of barley made a gallon and a half of whisky. In those days, the farmer made £12 per acre from growing the crop. It was then easy to work out that the revenue from taxation was between £300 and £400.

Today, with excise duty nearer to £5 per bottle and yields of grain far higher than seventy years ago, the 'tax take' for every acre of grain converted into whisky has been estimated to be in excess of £20,000 per acre.

Oatmeal millers

Traditionally, Scotland grew oats. The grain provided food, both for man and beast, the former through porridge and the latter as horse feed. The famous lexicographer, Dr Samuel Johnston, scoffed at oats, remarking that in Scotland they were used to feed men, while in England they were only used for horse feed, to which his Scottish scribe, Boswell, is reputed to have retorted, 'But look at the quality of the men in Scotland and the quality of horses in England!'

Even if the strength of the Scottish nation is reputedly built on the porridge-based diet, evidence from the past century is that a fair percentage of the oats in porridge comes from abroad. Back in 1927, Cupar NFU wanted all imported oats to be labelled as such. One member wanted the label 'on every bag of oatmeal'. A century ago, oats were the main cereal crop grown in Scotland, with barley – used in the making of whisky – running a close second. With oatmeal a staple of the farm worker's diet, there was, up until the early 1950s, a network of oatmeal salesmen travelling from farm to farm. In the early days, a pony and cart would supply the customers who received meal as part of their perquisites for working on the farm.

There were a number of millers in the local towns. Firms such as J. & T. Rodger, Cupar, were typical insofar as they milled the oats and then took the oatmeal around the farms, where they would dispense it to the workforce. In Auchtermuchty, Sam Anderson, the miller, had spent four years earlier in his life as an apprentice miller in Aberdeenshire.

The oats to be turned into meal were first dried on floors heated by corn husks. It was important to check regularly that the grain was not 'smoked'. The method of checking the temperature was to spit on the stone floor and then observe how quickly the spittle frizzled. Once dried, the grain was evenly fed between grinding stones set just far enough apart to produce the correct size of oatmeal. The miller's secret was in knowing how to set the wheels, the top one being on a spindle.

Also a factor was the variety of oats used as some of the early varieties of 'potato' oats, so-called because they were grown after a crop of potatoes, were thin-skinned. Some of the then newer varieties, such as Forward and Onward, were more difficult to mill and because of their thick skins they would leave less meal for the miller to sell.

When the meal was sold, it came in bolls, quarters, pecks, firlots or lippys; all of these were originally measures by volume, thus giving rise to differing weights being attributed to them.

In the old days, part of the minister's stipend, or wage, came in the form of oatmeal. For one East Neuk minister, the agreement was that he should get: 1 boll (62.5 kilos), 2 quarters (25 kilos), 1 peck (almost 4 kilos) and two lippys (approx 2 kilos). Later, when cash was preferred to actual oatmeal, Fiar's markets were held to establish values for grain and therefore determine the minister's pay.

Moving away from the small-scale miller and into the world of the international company: in 1947 the Scotts Porage Oats company bought the former flax factory at Uthrogle on the outskirts of Cupar. More recently, this facility – which is close to the former Cupar racecourse – has become part of Pepsico, a worldwide operator in branded food and drink products. Currently, as one of the two big processors of oats in Scotland, the company is buying in some 90,000 tonnes of oats annually, with 50% of the end product exported to more than seventy countries around the globe.

Potato trade

In the early part of the last century the majority of the potato crop was traded at Cupar market, although one participant recalls the Imperial Bar opposite the Corn Exchange being a popular trading venue. There, the Glaswegian ware merchants, or their local buying agents, would meet with farmers about to open up their potato pits.

Likewise, the seed traders would try and fill their order books by buying the varieties and grades of potatoes in demand. Often there would be a barter system in operation with sales of eating potatoes offset through new seed purchases. Prior to the identification of individual seed crops with specific farm numbers, it was possible for merchants to sell farmers back their own seed. This regularly happened after the consignment passed through the hands of a number of trading agents, who all took a financial cut.

The life of a trading merchant living on a narrow commission was never easy. A falling market could lead to the dreaded early-morning call from a buyer located in the big city markets indicating that the potatoes which had just arrived were 'not up to expectations'. Faced with the cost of a return journey or accepting a price cut, the latter option was often taken. Price fluctuations were always more severe with the early season crop and it was possible to lose large sums of money in a short time, if the market 'went wrong'.

In the early 1960s, Ladybank-based Stokes Bomford pioneered the pre-packing of potatoes, putting them in small 3 or 5 lb (1 or 2 kilo) bags that the housewife could easily pick up. The potatoes were washed before packing, marking the start of a move towards food that could be more easily handled and cooked. This trend brought more and more sophisticated equipment to the potato pack-house, and now electronic weighing machines price the end product down to the last penny. Almost all of the smaller packhouses have now gone and more than half the Scottish ware crop goes through four large packers, working for the supermarket trade.

The seed trade has also integrated and independent merchants have largely gone, to be replaced by companies that control varieties right from the breeding stage through to commercial production.

Seedsmen

Many improvements that followed the Agricultural Revolution were obvious even to those outside the industry. Improved breeds of sheep and cattle, drainage of fields and cultivation techniques were there for all to see. However, for the layman looking over the hedge or through the fence, grass is often just seen as grass. This undervalues the contribution of the improvers of the various species and varieties that feed and fatten the livestock.

It is also unfair on the network of seedsmen, who have a specialised knowledge of the characteristics of the various grasses and clovers and can use this knowledge to satisfy specific needs. In the last century, Cupar had three special-ised seed houses, supplying grass seed mixtures, seeds for a wide range of other forage crops and vegetable seeds for rural and urban gardens.

It was quite normal for a grass seed mixture to contain a dozen or more different types of grass and clover. Some, such as cocksfoot, start growing early in the spring and thus provide grazing in March and April. Others, such as the ryegrasses, are known for providing yield for farmers making hay. Then there are soil conditions to consider, with some grass species thriving on light land and others preferring stronger soils.

Farmers would then add some clover to their mix. This can not only be grazed, but with its nitrogen-fixing roots, it also acts as a fertiliser to neighbouring grasses. Again, there is a choice to be made with some such as Kent Wild White being more suitable for longer leys, while red clover from England helps bulk up a crop of hay as well as improving the eating quality.

After the required formula was agreed, seedsmen mixed all the ingredients on the floor of their premises. Then the staff, using wooden shovels, would ensure a thorough mix before bagging it up into sacks.

The seedsmen would also provide turnip and swede seed. In the early days, they would often enter a field of one of the more popular varieties and then select some superior plants. By keeping the seed from these and then multiplying it up, they would then be able to sell their own selection. In one example, Watts, the seedsmen from Cupar, sold what they described as 'Watts Excalibur turnip seeds'.

The seedsmen also sold a wide range of forage crop seed. In the early part of the century, tares and vetches were popular but these have now slipped out of commercial sight. Lucerne has had its followers, especially those farming on light land, but it has never achieved any significance in this country. Catch crops, such as mustard, are still occasionally seen. As it can be sown later, and grows quickly, a mustard crop often indicates that the original crop has failed.

Turnips are not the only root crop available to livestock producers wanting winter feed: some farmers, attracted to the big tonnages that mangolds could produce, grew a few acres of them. But this crop suffered as it did not lend itself easily to mechanical harvesting and anyone who has hand-pulled and loaded mangolds will remember the pain.

Co-operatives

One of the oldest sayings in the farming industry is that two farmers will only work together or co-operate in order to beat a third. Unlike our European neighbours, farmer co-operation has not been a dominant factor in the agricultural industry. Talk of co-operation increases during times of hardship and difficulty; equally, in times of financial security and wellbeing, the thought of co-operation slips down the agenda.

SAOS, the umbrella body for all Scottish farm co-ops, was established in 1905, but many other individual organisations were born in the hungry years that existed prior to that time. In Fife, the East of Scotland Agricultural

Co-operative Society was in operation in the late nine-teenth century, buying seed and fertiliser in bulk on behalf of its members.

In the hard post World War I days, another farmer co-op based in Fife made its first appearance. Central Farmers began life when four tenants of Wemyss Estates joined together to buy their fertilisers more cheaply. They succeeded and subsequently told their tale to the Thornton branch of the NFU. Then, in June 1924, in the Thornton Auction mart offices belonging to John Swan & Co., the West Fife Agricultural Trading Society was set up. For the first ten years of its life, this co-op worked on a commission basis, with only an office in Methil to mark its existence.

During this period, the co-op changed its name to Central Farmers. Like all successful co-operatives, Central Farmers was run by a small number of strong individuals. Archie Dryburgh came in as chairman in 1932 and at the same time Alex Mitchell took over the manager's post. These two were followed respectively by John Arbuckle and Andrew Smith, and as these latter two were still in post in 1982, not many organisations can point to having only two chairmen and two managers in a fifty-year span of working. In 1932, the co-op took over a former colliery on a 3-acre site in Methil to set up a manufacturing base for fertiliser. Within three years, this venture was paying divi-dends, with members sharing almost £60,000 in bonuses.

The co-op expanded into buying potatoes and grain from farmers, while at the same time widening its selling range with animal feed stuffs, grass seeds and requisites such as binder and baler twine.

At its height, the annual turnover of Central Farmers exceeded £20 million, but by the 1990s, the nature of farm trading had changed. No longer were sales representatives needed to go up farm roads – a costly way of conducting business. Imports of fertiliser undercut home-made versions and cereal and seed trades became more competitive. Central Farmers posted large annual losses. It closed its fertiliser works, stopped making animal feed and moved its base to Milnathort. By 2006, it had closed its doors, and after paying off shareholders, a small residual balance was paid into the Royal Scottish Agricultural Benevolent Institution.

There have been several other farm-based co-operatives in Fife in the past century. Many have drifted out of existence as events have changed. There used to be a number of rabbit clearance societies, but they all folded in the 1960s. Beef producers set up a co-operative specifically to market cattle reared on barley, but again this failed the test of time after many of the cattle became afflicted with liver complaints.

A new generation of co-operatives now exists. One example in Fife is the Tay Forth Machinery Ring. It only emerged in the past couple of decades, but this machinery and labour sharing organisation has helped farmers cut both their capital and running costs. Now it forms part of a national network of similar organisations.

Numerically there may be fewer co-ops than there were mid-century, but they are doing more business, with around half the total output from Scottish agriculture now reckoned to come through co-operative organisations.

National Farmers Union involvement in trade

Nowadays, the National Farmers Union of Scotland does not carry out any direct commodity trading, but back in the mid-1920s Anstruther NFU decided to buy binder twine on a wholesale basis for the benefit of members. The secretary was asked to write to the agents of Maypole Leaf,

John Bull and Buffalo brands to get quotations. In return, the binder twine makers wrote back, saying it would not be wise for the branch to book at present – it was possible the price may come down. Finally, it was agreed that the Bulldog Twine Company would get the order (for approximately 5 tons) at 51/- per cwt, or £2.52 per 50 kilos. This trade continued up until 1930, by which time Union members were receiving the benefits of co-operative buying from Central Farmers.

Chapter 25

Shows

ALTHOUGH far from its original purity of aim – the promotion of the science and husbandry of agriculture – the claim by the Highland Show to provide a big day out for the whole family has resonated with farming people the length and breadth of Scotland for the last century.

In the early 1900s, transport limited movement, but the 'Big Show' travelling around various areas of the country compensated for this. The attraction has long been a combination of education and entertainment, and that is why my early memories of attending the Highland always seem to relate to lengthy trips by car, where the 'Are we there yet?' question reached crescendo pitch.

This was followed by father stopping to talk at length to what seemed like everyone who came past; after an hour or two, the family split up, leaving father to his farming talk while mother and the offspring went walkabout. She must have steered a canny course through the showfield as we never seemed to come across the candyfloss stalls until the very last moments. By this time, there had been the inevitable loss of a brother, or more fortunately, little sister in the mêlée. A standstill order was placed on those remaining until the team was brought back up to full strength.

We always managed to see the livestock lines; we even

went to see the goats. In a serious, yet unproven allegation that has survived the past fifty years, my little sister claims that she was pushed into a bucket of water at this juncture. A dripping and noisy child brought this particular show visit to a quicker finish than normal.

Looking locally, in its 200-and-more-year history, the Royal Highland Show has only once been held in Fife. This momentous occasion was in 1912, when the premier agricultural show in Scotland was held on the outskirts of Cupar in the grounds of Kinloss Farm, a site which is now the base for the annual Fife Agricultural Show and which, in the late nineteenth century, was the original location of Cupar golf club.

While the RHS directors, under the presidency of Lord Ninian Crichton Stuart, from Falkland, supported the peripatetic show as the best method of ensuring support from all over Scotland, the Highland Show visit to Fife was unusual in that Cupar was one of the smallest venues ever chosen.

It may have been the county town, but it did not have the local population such as could be gathered from cities like Aberdeen, Inverness and Dumfries. With a limited rural hinterland and the small size of Cupar itself, the directors recognised there was a risk it would not garner sufficient support, both in potential spectators and in trade stands, thus leaving them with red ink on their balance sheet. Prior to the decision being made, the Highland Show directors did check that the site was in close proximity to a railway station. In those pre-road vehicle days, this was vital as livestock could then be brought from all over the UK. Cattle, horses and sheep were all walked from the railhead, the mile or so to the showground, through the streets of the market town to the 35-acre show site.

The Town Council of Cupar offered free water for the

livestock, and among other supporters, the Fife Foxhounds promised a £60 contribution to the show funds.

In those days, when the Highland Show travelled around Scotland, a great deal of equipment linked to the show had to be transported from site to site. The work of constructing the various pavilions and livestock pens was contracted out. Some four weeks before the event, the local paper reported a mile of perimeter wooden-slabbed fence – some 9 foot high – had already been constructed and work was proceeding on the main pavilions. It was reckoned that in those days, more than 1,000 tons of timber was required in setting up the show, between the needs of the perimeter fence and the largely wooden pavilions.

There had been a slight hiccup in the construction work with the discovery of three prehistoric cists (grave pits) on the site of one of the stands. However, work soon resumed on the 'large and handsome' red mahogany wood Directors' and Secretary's pavilion dominating the central show square and the reporter noted four weeks before the event that its construction was already well advanced.

Unlike today's shows, where there are numerous fast-food outlets, working from mobile sites, the main catering pavilion at the show was based in a large 120 × 56 foot tent. To keep the social classes apart, it had first- and second-class public rooms.

As befitted Scotland's largest outdoor annual event, there were major pavilions from Australia, Canada and Rhodesia (now Zimbabwe), where the benefits of farming on these Dominion lands were proffered to Scots who might be less than contented with their lot. It is interesting to note that with a very depressed agricultural industry at home in that pre-war period, the show reports indicate these foreign pavilions were busy.

There were also more than 200 trade stands and a special

demonstration area for 'mechanical equipment'. This development chimes with the founding articles of the Highland Society and the pledge that the Society would be at the forefront in new technology and husbandry.

This area featured a wide range of static petrol-powered engines puttering away gently, while demonstrating their ability to turn the wheels of mills, feed grinders, turnip cutters or early versions of milking machines. For a population used to hand labour, or at best, original horsepower, these machines offered a glimpse into the future, where machinery would increasingly take over the hard work on the farm. There were early tractors and also motorcars, many with exotic names such as Studebaker, which have now slipped into history. Others, such as the Scottish produced Albion, testify to this country's role in the industrial revolution.

Along the trade lines there was also more locally produced equipment, with several blacksmiths demonstrating homemade coup carts, field harrows and drill ploughs.

Wind turbines are now back in the news as potential providers of renewable energy, but almost a century ago, the Cupar Highland show had Coopers of Dunfermline demonstrating windmills for extracting water from wells for farms. They must have done so successfully as several old multi-vaned water windmills are still to be seen around the farms of North-East Fife.

A number of Cupar shopkeepers also took stand space, no doubt thinking of the great opportunity to expand their customer base. Firms such as the ironmonger Dott Thomson, that still exists to this day, were selling zinc buckets and hand tools, which were important items on farms in that period. J. Gilmour, the newsagent, promised those visiting the show, 'the latest news, both from the wider world and from the show field itself', while

Honeymoon, the saddler, exhibited his wares to an audience for whom the horse still played an important role in business and social life.

Telephone lines were routed to the show from the local network and arrangements were made for electrical power to be generated on site. These services, still in their infancy, had specialist workers contracted to be on site for the duration of the show.

The livestock pens were all under canvas, with individual horseboxes receiving particular commendation from the local journalist.

To this day, one of the highlights of Scotland's premier agricultural event is the Grand Parade of livestock. At Cupar, a grandstand with a capacity of 1,600 people was erected for the best viewing of this spectacle. Unfortunately the show was blighted by an occurrence outwith the control of the organisers. Foot and Mouth Disease had broken out in England just prior to the show and all the top English livestock, along with Scottish exhibitors who had travelled down to their premier show, the Royal, two weeks earlier, were stranded in the south because of livestock movement restrictions imposed by a government desperate to contain the infectious disease.

Despite this drawback, the show went ahead, with some 60,000 coming to see the cattle, sheep, pigs and horses under the judges' eye. With this level of support, the directors' fears were not realised and there was a small profit of £20 from takings of £3,579.

Unrelated to the show, but at the 'greeting meeting' of the Royal Highland and Agricultural Society that followed the Cupar show, the Earl of Stair suggested warships off the West Coast of Scotland should curtail their practice firing of artillery during the months of August and September. The good Earl claimed the discharging of the guns burst the rain clouds to the great disadvantage of agriculturalists. Directors of the Highland Society agreed unanimously to write to the Admiralty to express their concern.

Fife has never been renowned as a powerhouse for pedigree breeding. There have been, and still are, several notable breeders who have reached the top of the tree, but with its largely arable base, it does not have the concentration of livestock men that an area such as Aberdeenshire merits. As a result, all the top prizes in the Cupar Highland in 1912 went outwith the home territory, but it is interesting to note that several names which carried away the top championships tickets a century ago are still noted breeding families and farms today. In the Blackface sheep section, Matthew Hamilton of Woolfords took the championship and Elliot of Rawburn the North Country Cheviot championship.

Currently, the Highland Show has some 16 breeds of cattle and 23 of sheep, following the massive surge in imports of Continental breeds in the 1970s. The range of livestock at Cupar was more limited, with only the Aberdeen Angus and Shorthorn breeds representing the beef section and Ayrshire cattle the sole breed in the dairy section.

While the top Scottish farming show came only once to the area, Fife, like every other district of Scotland, has its own local shows that span the twentieth century.

Fife Agricultural Society started in 1888 with Capt. John Gilmour of Montrave as its first president. In 1904, it amalgamated with the Cupar and North Fife Agricultural Society, the Fife Farmers Club and the Fife Clydesdale Horse Society to become the top local show in the north of Fife. Three years later, it amalgamated with the Windygates Agricultural Society, with Sir John Gilmour still president.

In the East Neuk, the main show was the Colinsburgh show, which continued until 1954, when it linked up with the Fife Show under the banner of Fife Agricultural Association, incorporating the East Fife Agricultural Show and the Fife Agricultural Society. This move followed approaches by the Fife Agricultural Society, which had made a loss on its last show held in Kirkcaldy. 'Not a place to hold an agricultural show,' one director declared, with the benefit of hindsight after the event.

Following this loss-making show, in 1937, there was an appeal from the Fife Agricultural Show to the West of Fife Agricultural Society to amalgamate and put on a single show that would rotate around the county. There was support in the East, but those living in the West of Fife turned down the idea.

A decade later, the issue was complicated with the arrival of a new show based in the Leslie area. Again, those in the East approached the directors of the new show asking if they were interested in amalgamation and again, they were rejected. It was only later that the Leslie show linked up with the West Fife show, leaving the present-day position of two summer shows: one in the East of the county and one in the West.

It was the habit of these local shows to move around their locality until costs dictated a more permanent venue. Currently, the Fife Agricultural Association shows are based at Kinloss – very close to the site of the 1912 Highland show. This location turns around a decision made back in 1929 when the committee decided against it as a location as it was 'rather out of the way.' In those days Tarvit Farm was preferred because it could be accessed by special livestock trains, which were arranged by the railway company. The show has been on other sites around Cupar, with a post World War II show held at Wetlands, which are now the playing fields of Bell Baxter High School.

In a comment on the method of transport to the 1929 show, it was recorded that some 302 cars paid 2/-, or 10 pence, each at the gate, 32 motor bikes paid 1/-, or 5 pence, each and 19 pushbikes paid 6d, or 2.5 pence, each at the gate.

The history of local shows is as affected by animal disease and war as much as by the industry itself. After five cases of foot and mouth disease in Fife, including one at Muircambus, with a bought-in bullock, the 1931 Fife show was cancelled. To this day, there is the constant threat of disease hitting one part of the country and setting in place movement restrictions. Even when they are held, there have often been divisions in the livestock pens. For a long period, these related to tuberculin-tested cattle and non-tested. More recently, there have been splits in the sheep ranks, with breeds susceptible to an infectious disease – *Maedi Visna* – being kept apart from the rest.

In 1915 the directors wondered whether to proceed with the show in Thornton because of the war. It was decided that it would go ahead but no prize money would be handed out.

It was also decided there would be no industrial show, i.e. no best raspberry jam, scones, or shawl-making competitions as the directors believed, 'It would not be right to divert the ladies from knitting for the comfort of the troops.'

In the first post-war show, held in 1946, there was a problem sourcing sufficient tents as all the canvas was still being utilised in post-war efforts.

Entertainment at agricultural shows has varied over the years. Back in the 1930s, events such as pitching the sheaf took place, as well as competitions to see who could harness a horse in the shortest space of time.

Tractor driving skills and forklift handling are the modern equivalent. In 1953, Fife organised a 'musical chairs' event, with competitors jumping on and off tractors when the music stopped, but that was before the Health and Safety Executive existed!

At the 1953 show of the East of Fife Agricultural Society, held in Deer Park, Elie, the committee organised special events to commemorate the Coronation of Queen Elizabeth II. Among the attractions were the Dagenham Girl Pipers, who provided the main ring entertainment with a programme of largely Scottish melodies.

The programme for that particular show demonstrates the change that has occurred in the last half-century of farming. There were 36 trade stands, tractor suppliers, implement makers and power companies, but only four of these businesses survive into the present day.

Other shows and competitions

Despite the entertainment provided, the basic rule remains: shows have been, and continue to be, one of the main meeting events for those living in rural communities. Apart from the summer shows, there has always been a series of events during the winter months. Notable among these is the North of Fife Foal Show, which even to this day provides Clydesdale horse breeders with an opportunity to compete with their latest breeding.

Up until World War II, a stallion show was held on the Hood Park in Cupar. This covered both work and driving horses, and provided onlookers with the opportunity to nominate the preferred sires for their mares. This local show was a pale shadow of the main Clydesdale stallion show held at Scotstoun in Glasgow, which was the event of the year when horsepower was the only power. Competition classes with more than 100 entries were recorded

and breeders from all over the country flocked to the
event.

Arable farmers, bitten by the exhibition bug, have their
event in the autumn with a root and grain show. Samples
of wheat, barley, oats and potatoes compete and there are
also classes for turnips and Swedes. Some sections, such as
mangolds, have drifted off the competition scene, while
others, such as silage, replace them. Underlining the competitive instinct in the farming community, there were 82
entries in the turnip classes, 21 in the sugar beet section, 3
in the field cabbage class and 18 exhibits in the seed wheat
section at the 1929 show.

Even outwith the boundaries of the county, Fife men
have played a leading role in agricultural shows and events.
The Scottish National Fatstock Club was set up in 1896 and
runs to the present day. It was among the many enthusiasms
of Frank Christie of Dairsie Mains. He was also an exhibitor
of fatstock at many of the top winter fairs, such as Smithfield, and after considering Scotland should have its own
event, he went ahead and helped organise it. The first president of the Club was Sir John Gilmour of Montrave, Leven.

The competitive instinct in the farming community also
rises with annual ploughing matches. These have been
held throughout the century wherever land is under
cultivation.

In 1948, Bell Baxter Agricultural Discussion Society – a
local young farmers club – held their annual match at
Hilton of Carslogie, on the outskirts of Cupar, where some
45 tractor yokes and 14 pair of horse turned out to demonstrate their skills. A decade earlier, in the mid-1930s, the
Stratheden Ploughing match had 30 entries in the horse
section, while in that same era the Highlands of Fife
Ploughing Society, with a restricted entry from around the
Peat Inn area, attracted 29 pair of horse.

Although the numbers of horse competitors has gone from all but the National ploughing championships and a few specialist events, competitive ploughing matches continue. Cups and salvers are awarded to those showing the most skill. Not only are there overall prizes, but the ploughman with the straightest rig or the best entrances and exits, or 'ins and outs' as it is termed, go home with reward for their effort. At some ploughing matches even the very subjective, 'Best-looking Ploughman' receives an award, although this is often a token rather than a financial prize.

Even if the Highland Show has been permanently based on the outskirts of Edinburgh at the Ingliston showground for forty years, a day out there still features in most farmers' diaries. Before it settled down in Edinburgh and the four-day show changed its opening days from Monday to Thursday to the current Thursday to Sunday, the traditional 'workers' day' at the Highland was the Thursday. Buses would be organised from the rural areas and employers often stumped up the entrance fee as a small reward for their workforce.

When it came to their own attendance at the show, Fife farmers sometimes drove a hard bargain. In 1948, when the Highland was held at Riverside Park in Dundee, the local NFU in Fife made an agreement between Tay Ferries and the NFU to run a special boat at 6.30 a.m. in order to get a full day at the show.

A few years earlier, the same local branch of the Union attempted to hire a bus for the Highland Show being held in Inverness that year. However, they could not persuade the authorities to provide additional petrol coupons for the trip. The journey may have been deemed important for the farmers, but the Government considered fuel rationing even more so.

Chapter 26

Pests

FOR a small boy in the early 1950s, one of the joys of harvest time came at grain harvest, when the binder cut its way towards the centre of the field. Actually, the highlight came when the binder, working in ever-reducing circuits of the field, had concentrated the rabbit population to a point where they started to realise they had to escape from the unharvested section. Soon they were popping out of the standing crop and heading across open stubble towards the nearest cover.

With sticks to prevent any great escape, we would be stationed around the field ready to decimate the rabbit population. We were fleet of foot and they were cruelly exposed after a lifetime casually chomping their way through the crop in the hidden safety of the growing grain. Escapees were quickly despatched with a clout from the stick and a quick 'rabbit punch' to the back of their necks when they were held up by their back legs. Before the field was finished, there were pairs and pairs of rabbits hung on the fence by the gate, ready for the local game dealer to come along and provide us with some pocket money.

This action was some small revenge on a rabbit population that could, in those days, quite easily reduce income from the grain crop by 10% or even 20%, with field adjacent to woodland or on lighter land where burrows could be easily created. It is strange to think that following their introduction to this country in Norman times, landowners encouraged the rabbit population as an addition to their diet. In Scotland, a statute issued by James VI actually required all landowners to establish a rabbit warren on their lands. On a wave of popularity enhanced by their contribution to the food baskets, rabbits were a protected species right up until the end of the eighteenth century. By this time, their numbers had multiplied to such an extent that they were regarded as a pest and the Game Act of 1880 allowed tenant farmers to shoot them to protect their crops.

In general, rabbit numbers were kept down to reasonable numbers for the first half of the last century. Rabbit meat was still seen in most butchers' shops and trappers made a reasonable amount of money. In fact, there was still a shortage of rabbit meat in the country, as in 1930, the UK imported some £3 million rabbit carcases and skins from countries throughout the world.

In the later 1930s, when the problem became too severe, farmers started to use Cymag, a poisonous gas containing sodium cyanide, which was injected into the burrows. This brought a storm of protest in the early years of World War II as many thought this gassing reduced the amount of available food for the population. However, gassing and trapping continued throughout the war and even prisoners of war got into the rabbit-trapping act, with several inmates at Lathocker camp, near St Andrews, being censured for being out of the camp at night, catching rabbits.

In post-war Britain, a move to more humane methods of dealing with animals came into vogue and there were letters to the newspapers objecting to the traditional methods of killing rabbits. A plea went out to farmers to use humane traps and stop using snares. The humane trap saw the rabbit fall into a box below the ground, where it was held until the trapper despatched it. However, the majority of farmers were unconvinced about the effectiveness of such traps, with one expert stating, 'The only way you can get one rabbit in such a trap is actually to put it there yourself.'

By the 1950s, the consumption of rabbit meat fell and soon the whole countryside was overrun with the fast-breeding furry pests. By 1954, the Government entered the fray with the Pests Act, which stated that rabbits were to be cleared from Britain's land.

Rabbit Clearance societies were set up and they employed trappers. In 1958, there were four such clearance schemes in North-East Fife. There were always one or two landowners who would not join such organisations, though, and their effectiveness was severely undermined by the gaps in coverage. In the late 1950s, there was a discussion about making all these rabbit schemes compulsory, but that came to nothing.

At Union meetings, there were several debates over landowners who were not keeping down the rabbit population. Allegations were made that some preferred to allow rabbits to multiply so that they could sell them. Favourite targets for criticism over a failure to keep rabbits under control were the Railway Authorities, with rabbits enjoying the built-up rail embankments as ideal bases for their warrens. Trappers were also accused of not trying to eliminate the breed because this would stifle their own future income.

In the mid-1950s, another solution to the rabbit problem came along. Rabbits in the south of England were reported to be suffering and dying in large numbers as a result of a viral disease called myxomatosis. By 1958, the disease had arrived in Scotland, leaving infected rabbits with ghastly swollen eyes, sitting around fields and country roads. These scenes of dead and dying rabbits also ensured rabbit meat was no longer regarded as a staple food.

It was thought the disease would wipe out the rabbit stock from the country, but a few survived, and within a decade, the rabbit population was estimated to be back to previous levels. Possibly linked to their survival was a change in habit, as rabbits started living above ground, thus reducing the chances of infection from the disease-carrying fleas.

To this day, the rabbit population rises and falls. There is a little trapping and shooting, but carcase and pelt prices do not encourage this. Also, especially in late summer, there are occasional outbreaks of myxomatosis, but there is always a sufficient number of survivors to ensure a rabbit problem the following year.

Crows and rooks

Generations of youngsters brought up on farms through-out the world will testify that their first working chore is to keep pests and vermin from eating the family crops. From the beginning of time, or at least from the start of growing crops, small boys were sent out to the fields to scare away birds. As an alternative to school, this had its attractions, being a particular form of homework not given to those living in the towns.

In the early years of the century, a common method of bird scaring was a wooden rattle that provided a rat-a-tat-tat as it revolved. Just how this 'rackety', as it was called locally, was later converted into a popular weapon in the hands of football supporters, it is difficult to tell.

In addition to the noise deterrent, it was traditionally popular to create scarecrows purporting to resemble humans. Old clothes stuffed full of straw gave a passing

likeness provided, that is, the birds that required scaring
were somewhat short-sighted. The tradition of scarecrow-
making continues to this day, albeit with more sophistica-
tion and style.

Thankfully, by the time my bird scaring days arrived,
more sophisticated forms of bird deterrent were available.
There were rope bangers, which, when lit, saw the rope
slowly burn down to an entwined explosive charge. A
rapid bang followed, which scared the crows and ensured
the crops were safe, at least in the 10-metre radius around
the rope. The government also provided its own bangs,
with a subsidy for ammunition until 1950. Even after this
financial support had been removed, some farmers wanted
to see neighbours with rookeries fined. However, as one
opponent of the scheme remarked, 'The crows eat the
wireworm pests in the cereal crops.'

If we were not scaring crows away from open feeding
troughs around the hen houses, or new-sown seedbeds or
even corners of nearly ripe grain that had lodged and lain
flat, then we went for more direct action. The crow cage,
or Larsen trap, is a wire cage with a hole in the roof
netting, through which the unsuspecting crow falls. It
cannot take the return trip through the small trap door: its
wings must be at full reach to fly upwards. The bird is thus
trapped, leaving only the humane despatch for this particular farming pest. To prevent this killing of mature birds,
one of the options for farmers in the arable and sheep-
stocked areas was the spring shooting of rookeries. Just
after the birds were hatched, shoots would be organised,
and the nests targeted.

In a small measure of revenge for this type of attack,
the guns would invariably return home adorned with a
combination of bird mess, twigs from the nests and even
egg yolk.

Pigeons

Somewhere in pigeon heaven, there is someone deciding that farmers should grow crops of which pigeons are particularly fond. In the early years of the century, the main target for the pigeons was the turnip crop which they always attacked when the seedlings were at their most succulent stage. Then, with an international shortage of vegetable oils, farmers were encouraged to start growing oil seed rape in the early 1970s. This crop has the added advantage, as far as pigeons are concerned, that the majority of it is sown in the autumn. This provides winter grazing for the pigeons.

In Fife, after the demise of the sugar beet factory, field-scale pea crops were grown. Again, this crop is at its most vulnerable not long after the seed pops its head through the ground. The introduction of field-scale Brassicas also provided a boost for pigeon fortunes, as crops of cauliflower, Brussels sprouts and broccoli are high on any pigeon's must-eat list.

Cereal crops do not miss the pigeon radar and flocks can descend on crops of barley and oats, especially at seedtime and harvest. The old saying for those sowing grain was: 'One for the pigeon, one for the crow, one to wither and one to grow'.

In pre-weedkiller days, pigeons were also seen as an ally. In 1930, a plea was made at Anstruther NFU branch for 'fair play for pigeons' on the basis that they also ate a large quantity of weed seeds, especially charlock.

In 1937, the same Union branch took issue with a Private Member's Bill going through Parliament, which was sponsored by the local MP, Mr Henderson Stewart. If it had come into law, the Bill would have made it illegal to shoot sparrows, crows and pigeons. In a strong response, it was agreed that the Secretary should see Mr Stewart about this outrageous proposition.

The main control mechanism for pigeons has been through shoots, on an organised basis or by involving local enthusiasts. There was a fashion in the 1960s and 1970s for bangers operated by gas to scare the pigeons off the crops. The mechanism was complex and involved water dripping onto calcium carbide, which then produced a gas. When the pressure of this gas built up, it was ignited and caused an explosion. That scared the birds away for at least five minutes, perhaps less in hungry times. However, it also aggravated all those within earshot of the gas gun because there was no mechanism for switching it off at night and it banged away for 24 hours per day.

During World War II, the Agricultural Executive Committee had a special sub-committee, whose remit was specifically to deal with any pests that reduced food production. One of their first moves was to organise weekly shoots, particularly against crows and pigeons. The shooting of pigeons and other pests around farm crops was supported by the Government right up until the late 1950s, with grants available for cartridges.

Rats

Brian Douglas is now grown up and working with motor-car giants, BMW, as an electronics consultant. This is a long way away from his upbringing on the farm, where his father was foreman. He took part in the games played around the farm steading. However, the occasion I remember vividly relates to something he did, which I could not do then and definitely could not do now. There was a stack of hay bales behind which came the squeaks of rodents. Inquisitively, Brian stuck his hand between two bales and then pulled out a rat by its tail. The rat hung upside down. No, the rat did not: it desperately tried to climb up its own tail to sink its sharp teeth into its captor. Sensing he might actually be bitten, Brian released it, but not before one or two of his friends had picked up shovels and other weapons of murder.

Rats have always been part and parcel of farming life. In the last years of the nineteenth century, a proprietary rat poison was advertised in the *Scottish Farmer*. The farmer who had used the poison claimed that 'it had left 150 rats lying about'.

A couple of decades later, in the 1920s, out in the potato field my grandfather kicked over a tuft of old grass to check the picker had gathered every last potato. He did not get any spuds. Instead, a rat sheltering under the clump latched onto his boot and then, seeking shelter, ran up his trouser leg. Thankfully, or this tale might not have been told, he grabbed the rat before it reached his reproductive parts and squeezed it to death. That is one reason why farm workers often wore nicky-tams – string tied round the leg below the knees – with sacking or canvas wrapped around the bottom of their trousers and tied tightly.

There were always rats around the farms. In city life, it is reckoned the nearest rat is less than 10 metres away, but anyone working with pigs or poultry during the early days of the last century would have been much closer to this particular pest.

A combination of easy-access feed troughs, readily available water and shelter would ensure that rats' noses would poke out from the pigs' crays (sties) and from under the hen houses. Mostly, the rats scuttled about in the darkness and they would rush for cover when an unwary person crossed an unlit close or first entered the stable.

Whenever there was a major upheaval in the farms, the rats would appear. The emptying of the cattle courts or reids would bring them out briefly, blinking into the daylight, before they headed for the nearest bolthole. It was possibly at threshing that there occurred the greatest rat-induced mayhem as they were unceremoniously turned out of their homes by the removal of the stacks. Often the mill men had small terrier dogs to deal with them, and on other occasions, a wire net fence would be put up around the stack to prevent the rats making their escape.

In 1941, with emphasis on not losing any home-produced grain, the Agricultural Executive Committee asked that all travelling mills had two rolls of netting, of no more than ½-in (1 cm) mesh. These would encircle stacks and catch rats as the stacks were threshed. Also, during World War II, some of the Women's Land Army volunteers were specially trained to deal with this very particular enemy. Often they worked with ferrets to flush out rats from their nests in the stacks and lofts. During that period, the Union also became involved and it was suggested at Cupar branch that something along the lines of a 'rat week' could be organised for an all-out attack on the rodents, but this proposal was not supported by the Cupar branch and it was left to individual farmers to get rid of their rats.

Geese

It is a cold October morning and potato pickers are waiting for the digger coming up the drills. Overhead, skeins of geese break the skyline and as they pass, their loud, distinctive honking fills the air. That almost clichéd scene is etched in my mind. It haunts me every autumn as the geese fly overhead, their magnificent 'V' formations helping them in their long migratory journeys. And yet I was brought up with little love for these large and amazing birds. They caused my father and many other farmers a great deal of wrath as they puddled about and grazed upon young shoots in the wheat fields.

My brothers and I were sent out with polythene bags

tied to posts. Placed strategically around the field, they were supposed to stop this theft and plunder. One week later, we returned to see how effective these flapping polythene sacks had been. A circle with a diameter of no more than 5 yards surrounded each post and flapping bag; the rest of the crop was well grazed.

The more effective method of control was always shooting. There used to be demand for geese, but this tailed off as other sources of meat became less expensive. Soon, the wildfowlers would only take one or two geese for themselves as the bird was always considered to have rather fatty meat.

Nowadays, geese are protected under legislation and shooting is only allowed within a short window in the calendar. The problem for farmers today, however, is that the goose population is increasing and to that can be added the concern about the numbers of geese that have given up their migratory habits. Now, many simply live all year round in the estuaries and lochs in Scotland. Like some of the human race, they have opted for the easier life.

Chapter 27

Legislation

A LIFETIME in farming and the reporting of farming events has provided many moments for discussion and dispute.

As a very naïve young boy, I wandered along beside the gang of men singling sugar beet on my father's farm. I had proudly told them that at the weekend I had seen a television at my grandmother's house. This was the first occasion when I had actually seen the now ubiquitous TV and I was keen to relay my tale.

Soon however, a discussion broke out between two of the singling squad. One maintained that the best TV pictures came from sets powered by gas, while the other held strongly to the view that the 'paraffin powered' TV would prove to be the best. Back and forth the argument went, just as their hoes clacked to and fro. I really couldn't make up my juvenile mind about all this; there seemed to be merits on both sides. There was only one solution and that was to ask my mother about which view was right when I went for lunch. She soon put me right on that particular issue.

Since then, I have taken part in, and reported on the many, many issues that swirl round the farming political pool. I would report on NFU meetings, where voices would rise and tempers fray. The issue was likely to be related to some obscure piece of legislation, which would impact mightily on the industry.

It was always a little strange to find the main contenders later having a quiet drink together in the bar. As a reporter, I have also been outside the main discussion chamber, where the European Union Council of Ministers was framing out their latest piece of bureaucracy. The press would be excluded and confined to a small, airless room, into which at odd intervals, the ministers would send their minions to brief journalists on how the discussions were going. As it turned out, these 'spin sessions', which were non-attributable, were invariably wrong.

However, despite this history of reporting on the issues affecting the farming industry, I have never heard any debate, or even opinion, on which pieces of legislation have most influenced farming in this country. Among the older generation, there would be supporters for the Agricultural Marketing Act of 1931, which brought some regulation into a very depressed market. Production of commodities such as milk and potatoes was regulated for the first time ever. It also encouraged co-operative marketing.

The effect of this Act may not have provided an immediate solution to the financial plight of farming that year, but it did give a long-term solution. While other Marketing Boards, such as the ones for hops and tomatoes, soon faded away, the Potato and Milk Boards lasted more than half a century and the Wool Board exists to this day.

The Act followed a decade when farm gate prices had been low, with the breaking point coming for farmers following the 1929 harvest. In March 1930, the local NFU branch organised a meeting in the Hood Park in Cupar. There, a crowd of 5,000 farmers, farm workers and local politicians heard calls for the implementation of an import tax to curb the inward flow of food. My grandfather, John Arbuckle of Luthrie, Cupar, spoke of the tons of home-produced food going to waste. He was supported by John Paul, the farm grieve at Kinnear Farm, Wormit, who said that the low prices brought about by imported food were endangering the livelihoods of farm workers. Incidentally, as the local paper reported, this was the first occasion when electrical amplification of the voice was used. Nowadays, that device is called a loudspeaker.

Other farmers would point to the 1947 Agricultural Act with its guaranteed prices as the most significant piece of legislation. Brought in by the radical post-war Attlee government, this was the first comprehensive piece of legislation for the industry. The Act guaranteed prices and introduced Deficiency Payments, which came into play when market prices fell below agreed levels. These payment levels were determined by the Annual Price Review; the major political event of the year for Union leaders for the following thirty years. The 1947 Act also introduced grants for liming, drainage, fencing, new machinery, buildings and livestock improvement. For farmers still with the financial scars of the late 1920s and early 1930s, this provided a backbone for production. Unfortunately, it also brought with it a tag of 'feather bedded farmers' as those working in other industries looked on enviously at agriculture and the subsidies it would be getting.

The younger generation would likely select the UK's entry into the European Union as the most significant move in the industry over the past 100 years. After years on the sidelines, during which time Germany and France took up pole positions within the new Community, Britain took the plunge into Europe in 1973. This heralded a period of rapidly rising prices and in the first few years within the European Union, farmers in this country benefited mightily from the Common Agricultural Policy (CAP). This policy had been framed in the hungry post-war years and the main agenda was to ensure a freedom from hunger. Farmers rapidly responded to a policy aimed at growing more cereals, producing more butter, and even, on the Continent, bumping up production of second-grade tobacco and increasing production of industrial-quality wine.

By the late 1970s, newspaper reports were full of articles on wine lakes, butter mountains and bulging grain stores, the operators of all these storage facilities making equally big mountains of money. The brake was put on over-production by taking land out of production, in a process called 'Set Aside'. This non-farming action runs counter to the natural instincts of the farmer, but this sense was dulled by promises of payments for a policy of non-production.

With some sectors of farming, such as grain and milk, receiving support and others, such as potatoes, pigs and poultry, outwith subsidy regimes, the CAP also helped sound the death knell of the traditional mixed farm. Perhaps the biggest negative legacy of the well-intentioned CAP will be that it took farmers away from meeting the demands of the marketplace and into production for production's sake.

While these three transforming pieces of legislation could be considered as having the most effect on the farming industry in the past century, others were seen as threatening. Back in 1911, the National Insurance Act was brought in by the then Prime Minister, Lloyd George. This required all employers and employees to pay the Government a sum of money as insurance against illness or unemployment. In today's terms, the money was not vast, being 4 old pennies, or 2 new pence from the employer, and 3 old pennies, or 1 new penny, from the employee per week.

The St Andrews and East Fife Farmers Club discussed

this contributory scheme at a meeting in the Corn Exchange in Cupar and 99% decided they were against it. Many also stated they would not pay.

Throughout the past century there have also been several pieces of legislation on farm tenancies. As has been seen, this was vitally important at the beginning of the century, when almost all farming in Fife was conducted with tenanted farms. Nowadays, with less than one-third of the land tenanted, legislation relating to the long-term letting of land is slightly less important. More fascinating are the terms laid out by landlords in letting their farms. Tenancies were quite specific about how the farm had to be handled. Rotations were spelled out in the legal documents.

Most rotations in Fife tenancies were based on the six-year roll of crops, called the 'Sair Six' of oats, potatoes, wheat, turnips, barley and hay, or in the local dialect, 'Oats, poats, wheat, neeps, barley and hay'. Often, the tenancy document also stipulated straw could not be sold off the farm as the landlord saw this as selling part of the 'goodness' of his asset. Neither could dung, the produce of livestock production, be cashed by selling it off the farm.

Somewhat surprisingly, given its widespread acceptance today, the use of nitrogenous fertiliser was not allowed on some tenanted farms. The only case where such a restriction now exists is where Sites of Special Scientific Interest are located. Other conditions included allowing the local Fox Hunt to operate without hindrance over the land. Those landlords who enjoyed shooting also put in clauses preventing the killing of game. Thus, farmers could only kill pests such as rabbits and pigeons that were attacking their crops, but game birds reared for shooting were allowed over the cropped land. This latter clause often caused ructions between the tenant farmer and the estate gamekeepers, who were alleged to have kicked holes in the rabbit netting protecting the crops so that the pheasants could feed. Other clauses in farm leases were individual to certain estates. One famous estate outwith Fife, Panmure Estate, Carnoustie insisted all the doors and guttering on the farm steadings were painted a certain colour: elephant grey.

Chapter 28

Organisations

As we sat around the kitchen table on a Sunday morning, my father would bring out his diary and announce that he would be away for several days in the coming week. By this time I was working with him on the farm, having decided that farming was deep in my blood. With father, it was a case of increasing absences from the farm as he took on more and more public work with organisations linked to agriculture, marketing boards, farm co-operatives and agricultural research institutes.

As any 20-something-year-old son of a farmer will testify, these absences were not regretted. On the contrary, they were welcomed as they provided an opportunity to make my own decisions on the farm work, or, as one older and wiser head commented, 'The opportunity to make my own mistakes.'

I confess I was lucky that my father did so much public work. Many of my youthful colleagues within the farming community were caught in a business where two or even three generations worked together. Sometimes that type of operation works, but it can also have a crushing effect on any weak link.

At the beginning of the twentieth century, there were few reasons for farmers to go away from their farms. Apart from attending the local market to buy and sell stock, and possibly to slake their thirst while sealing the deal, most of the farmer's time was spent on the farm. There is a massive difference today, where attending meetings is seen as part of the business of running a successful farm. These can be meetings where advisors provide advice on the latest husbandry techniques, or farming leaders head discussions on the politics of the industry. And that's before the official meetings finish and the real debate continues in the bar.

The past century was the first in agricultural history in which farming organisations and pressure groups brought about change. The Highland Society of Scotland may have had its roots further back in history but its main roles were in husbandry improvement and education in agriculture. Political pressure might have been brought on the government of the day by its distinguished directors, but it would have been done discreetly and not by public outcry.

The Scottish Chamber of Agriculture was established in 1864 and the original intent of this organisation was the reformation of the law relating to game. However, it soon developed into a body with a wider voice on farming and rural matters. In doing so, an eclectic membership that included farm tenants, landlords, factors and lawyers was gathered. Despite branches throughout Scotland, including one in Fife, membership of the SCA never rose above 1,000, and in 1939 it ceased to be after amalgamating with the National Farmers Union of Scotland.

Milk producers in the West of Scotland were more organised than their arable cousins in the East in the setting up of representative organisations. There was the Federation of Dairy Farmers Association of Scotland, whose sole purpose was to fight for better markets for milk. With the migration eastwards in the early years, they brought with them a background of collective strength, helping to secure better financial returns.

It was therefore not surprising that by 1913 meetings

were held to set up the National Farmers Union of Scotland, with one of the primary aims of the new organisation being an increase in the price of milk. Locally, Cupar branch was established in 1917, with David Lees of Pitscottie as the first chairman. Five years later, on 5 May 1922, Anstruther branch was set up, with Henry Watson of Drumrack, Anstruther presiding over a meeting of fifty farmers.

However, the early years of the Union were not always marked by success as the Cupar branch closed its doors in September 1930. The chairman, John Arbuckle of Luthrie, claimed arable farmers were subsidising dairy and sheep farmers in the West of Scotland. He had tried to change the subscription system but had been voted down at national meetings of the Union and saw no way out, but to disband. Little more than three years later, Cupar branch re-formed and both branches continued to operate until the mid-1970s, when they were amalgamated into the North-East Fife branch of NFUS.

Farm workers actually had their own Union before the NFUS was established, as the Scottish Farm Servants Union came into being in 1912. The driving force in this organisation was Joseph Duncan, the son of a farm worker in Aberdeenshire. He had been, like many other office bearers in the Farm Servants Union, involved in other trade union work. In his case, he had the grand title of general secretary of the Scottish Steam Vessels Enginemen's and Firemen's Union.

Under Duncan's feisty leadership, the Farm Servants Union established 150 branches throughout Scotland, with more than 12,000 members within the next twenty-four months. This impressive figure more than tripled by the end of World War I, its membership strength lying in the East of Scotland, from Aberdeenshire through Fife to the Borders.

This strength was echoed in the two strikes it organised, the first in Ross-shire in 1922, and a year later in East Lothian, where more than 1,000 men withdrew their labour. In both cases, the strike was called as a protest against the reduction in wages being imposed by farmers, who themselves were receiving less for their produce in post-World War I years. But the work of the Farm Servants Union was continually hampered by the movement of men; no sooner had a branch been set up than term-time came around and men moved to another farm. This had a major disruptive effect on established branches, which often saw office bearers and members move to a different area.

A national minimum wage for farm workers had been introduced in 1917, but the Depression following World War I saw any wage agreements torn up. Cleverly, Joe Duncan never argued for a minimum wage in those distressed times, fighting instead for a right of collective bargaining. One issue the FSU raised unsuccessfully was the removal of Saturday afternoon from the working week. This was rejected by farmers, who calculated that 52 Saturday afternoons equalled 26 working days per year; a major reduction in employment. In 1932, the Farm Servants Union amalgamated with the Transport and General Workers Union. Up until the mid-1960s, some of the men on the farm continued to pay their weekly subs to the Union, which were collected by a local activist who cycled from farm to farm.

In 1937, the Government established the Agricultural Wages Board with a responsibility for setting minimum wages and work conditions. One year later, almost one-sixth of farms inspected showed cases where employers were paying below the legal amounts. Despite the arrival of a National Minimum Wage in 1997, the

Agricultural Wages Board exists to this day and continues to set rates of pay and work conditions.

Education

In the early part of the last century, students interested in a career in agriculture would head off either to one of the three agricultural colleges in Aberdeen, Edinburgh and Ayr, or to one of the universities – Aberdeen, Edinburgh or Glasgow – where a degree in agriculture could be taken.

The colleges had been established with 'whisky money' that had come from additional duties imposed on spirits and beers. This cash was originally intended for police purposes, but the Government channelled the money into agricultural education and research instead. This left the local Council in Fife to provide education at a lower, more practical and less theoretical level. Fife was the first in Scotland to run courses on poultry husbandry. The lecturer in charge of this, Miss Kinross, reported attendance at these training courses was exceptionally good, as keeping poultry was an integral part of farming in those days.

During the 1920s, as part of a packet of measures aimed at improving practical skills, government money was made available to the local authority for technical colleges. In Fife, some £23,000 had been allocated for this purpose. A mining school had already taken a quarter of the budget before a proposal came from the Fife Education Authority for a small farm to be bought.

Unfortunately, this proposal, which would have cost about £5,000, tore the local NFU apart. It was described by one member as 'uncalled for and unnecessary' while another noted that ploughmen would rather 'play football than attend an experimental farm.' The chairman set the tone of the discussion by stating that he had started life as a ploughboy and had he not saved up £100 to go to college, he would have, 'ended up as a clodhopper'.

Taking the opposing side, one speaker said that if he had not received education, he would have been driving a horse and plough, but instead he had risen to become town clerk of Cupar. Another supportive contribution was to the effect that colleges were there, 'but they were not for the masses'. One irate member pointed out that the same authority was spending more on training shopkeepers than on educating farm workers.

As the local reporter noted, there was by this time, 'quite a stir', especially as one of those not wishing the farm to be established was revealed to have a conflict of interest – he was on the Board of the Edinburgh and East of Scotland College. The proposal died, and another forty years were to pass before a farm was bought for practical education in Fife.

In other NFU debates, the attitudes of farmers to the schooling of children in rural areas were often narrow and self-interested. In 1920, one stated, 'School masters have informed me that children return to school in much better health and more able for their lessons after a potato gathering than they were before it.'

Later in the same decade, the local branch of the NFU unanimously supported a motion that called for children in rural areas to complete their education at their local village school. One member stated he did not want children to move out of rural areas, which would 'swell the numbers of unemployed' in the towns and cities.

Despite these attitudes, agricultural education advanced. In the 1920s and 1930s, agricultural courses run by the County Council linked up with the Edinburgh and East of Scotland College, and there was an educational base in

Cupar, at Bell Baxter School, where 'continuation' night classes were held. A similar facility was established in the grounds of Madras College in St Andrews. In the early 1930s it was not unknown for students to turn up on horseback at these evening classes – rural transport was not well developed in those days.

Ironically, the depressed state of agriculture helped the local education initiative as money could not be found to send youngsters to Edinburgh or Aberdeen, but they could afford the 5/-, or 25 pence, for the thirty-week night class. As a result, it was reported that evening classes were 'fully subscribed'.

Dr John Wilson, by his official title, or 'Wearie Willie' as he was affectionately named by students, was in charge of both locations and in 1942 it was claimed that Fife had the best 'continuation' classes in the country. Further accolades arrived in 1948 when these classes produced the first 'night school' students to gain National Diploma in Agriculture awards.

This era saw a push for more food being produced and support was given to agriculture to hit that target, so it was not too surprising when Anstruther branch of the NFU wanted the ancient University of St Andrews to become involved in running classes for farm students. This initiative was not taken up and so farming leaders then pushed for a specialist centre in Cupar. In 1957, Elmwood Training Centre, Cupar was established by Fife Education Committee in a large house on the outskirts of town. Under the control of David Bett, two years later it became a college and then in 1967, the first moves were made to purchase Springfield Hospital Home farm.

Success in this purchase saw a generation of agricultural students trained in practical skills and this continues to the current day, although much of the farmland is now a golf course and the college has recently gained an international reputation for training green keepers.

Young Farmers

The 'continuation' classes held at Bell Baxter in the 1930s did not just produce students with diplomas and certificates, it also helped organise a 'get-together' club for students, where those with similar interests in farming could discuss issues and have debates.

This grew into one of the first young farmers' clubs in Scotland. In fact, Bell Baxter Agricultural Discussion Society existed before the national body was set up. The BBADS title led to an early dispute between the local club and the national organisation (who wanted all affiliated clubs to be called Junior Agricultural Clubs). Faced with this proposal, the leaders of BBADS decided they would stay outwith the national organisation. They had a very strong financial base, also a large membership and it was no surprise when the Scottish Association of Young Farmers' Clubs relented. This leaves BBADS as one of the few junior agricultural clubs without acknowledging the fact in their title.

In April 1944, Anstruther branch of NFU suggested that a Young Farmers Club be formed in the Colinsburgh area and a month later, at the 30 May NFU meeting, it was agreed that East Fife Junior Agricultural Club be set up. Since these days, both clubs have contributed to the national scene, winning stock judging competitions at the highest levels and providing office bearers for the top places on the national scene. Many representatives from this area have travelled afar on scholarships and a number have moved on to take up leading roles in national farming organisations.

Along the way, these same young farmers' clubs also

gained a reputation as a rural marriage bureau. My only comment is that over 90% of the chairmen/women and club secretaries who held office in the first fifty years of Bell Baxter ADS's existence married someone involved in agriculture.

Chapter 29

The Future

AS I sit watching the slanting autumn sun, the geese have arrived and are squawking contentedly on a stubble field. In recent times, this field would have already been under the plough, but harvest this year has been slow and difficult.

There is also less incentive to sow cereals for next year as costs of fuel and fertiliser have risen almost as dramatically as cereal prices have collapsed. I look across the river Tay to the hills and glens of Perth and Angus. These are some of the traditional livestock breeding grounds, where hill farmers produce the stock for fattening on lower land. Such stock farms are slipping out of production: there is now no subsidy linking the production of lambs or store cattle to farmers' income. It has been estimated that one-quarter of the £400 million received annually by Scottish farmers now goes to those wearing the slippers of retirement instead of working wellie boots.

For the past 150 years, international trade has affected UK farmers. The only difference today is that reaction to events and world circumstances comes more quickly, and more unexpectedly. Inevitably, there are calls for 'national food security' – a method of protecting production of food within these shores. The call went up in the first decade of the last century, when wheat, lamb, beef and wool flooded in from across the globe. It was repeated after World War I, when the farming community reminded everyone how hungry they were in the latter years of the war. But it was a hopeless argument in an era of economic depression when cheap food was important to the nation.

It was not until World War II that the issue of national food security returned to the agenda. Years of virtual blockade, with merchant shipping being targeted by the enemy, made home food production vital. In 1944, one of the first-ever debates organised by the East Fife Young Farmers Club was whether national food security should be a national priority. Incidentally, the second debate that evening was, 'whether farmers' wives should be country-bred'. In both cases, the 'Ayes' had victory.

While it cannot be claimed that debate in Fife influenced politicians, the post-war government did set up schemes to support agriculture and boost food production. Every aspect of the industry, from ploughing up extra land to liming and subsidies for the end product, was included in the 1947 Act. There were concerns, but these were related to anything that might negatively impact on the expansionist programme. One example came in 1950, when the NFUS complained about, 'the pace of encroachment onto fertile land with the building of new towns, roads and aerodromes.'

Throughout the 1960s and 1970s, the farming industry remained expansionist. Politicians were still writing policy papers, with headings such as, 'Food From Our Own Resources', and in 1967, the local college advisor in Fife encouraged those farming on the shorelines to look at 'reclaiming land from the sea'.

Today's politicians are just as unlikely to support a food import barrier as their predecessors were in the 1920s. The consuming public has become accustomed to cheap food and that is unlikely to change.

Bureaucracy

A recent survey found that farmers are now office-bound for the equivalent of one day every week. If their predecessors had looked over their shoulders, they would stare in astonishment at the cattle passports now required for every bovine beast, the codes of practice for this, that and every other thing, and the legislation surrounding employment. Had he come from the 1920s, though, he might have recalled that even in those relatively paper-free days, the NFU still felt it necessary to write to the Department of Agriculture about the 'crushing' level of bureaucracy in farming, with the 'clerical burden retarding the efforts of all in the industry'.

Entry into agriculture

Many of those I interviewed, who were enjoying their later years of life, referred to their start in farming in the tenanted sector. Back in 1918, it was reckoned a tenant farmer only needed about £1,000 capital for every 50 acres of land. That sum included all machinery and livestock costs; it was also the acreage reckoned to be what one family could work. Further back, in 1893, a report into agricultural labour reported, 'the young ploughman who takes care of his wages and does not fritter them away in foolish ways has every opportunity of saving in ten years sufficient for him to become a farmer.' This opportunity was seen as the first rung on the farming ladder – a ladder that could take the willing, keen and able up into land ownership and expansion of the acreage farmed.

Now, with land prices over £10,000 per hectare and costs of machinery such as combines hitting the £250,000 mark, there is a view that all the rungs of the ladder have been removed. Or, as one interviewee more used to a free and untrammelled way of life mused, 'The Health and

Safety Executive have removed the farming ladder on grounds of financial insecurity.'

The contrary view is that opportunities have never been better for those wishing to enter the industry. Land ownership is a powerful emotion, but the absence of it does not disqualify those wanting to run their own businesses. There is more fluidity in the industry, with machinery contractors and livestock share-farming opportunities, for example. Increasingly diverse food demands from consumers bring opportunities for those willing to try something different. It is noticeable that the sectors of farming that survived – potatoes, soft fruit and poultry – outwith the protective umbrella of the European Union Common Agriculture Policy are currently among the most market-orientated and successful. Subsidised sectors will take time to regain all of their cutting market edge.

Acknowledgements

Over the past fifty years, I have read many books relating to agriculture in this country. As any reader will testify, some of the thoughts and themes are absorbed through the brain cells. However, there are several books and pamphlets that specifically helped in the writing of this book and I acknowledge them here. They are:

An Overview of Agriculture in the County of Fife by the Reverend John Thomson.

Archives of the *Fife Herald*, held in Fife Council Library, Cupar.

'Grain Production': a pamphlet produced by the Edinburgh & East of Scotland College of Agriculture in 1950.

Highland Agricultural Society Annual Year Book, 1912 for specific reference to the Highland Show held in Cupar.

Lairds and Farmers in North Fife by Robert W. MacLeod: his labour of love details farms and farmers in North-East Fife.

Ring of Memories by John Thomson: a comprehensive history of livestock auctioneering in Scotland.

The Agricultural Labourer Royal Commission: a report by Hunter Pringle published in 1891.

In addition to the above, there were a large number of kind people who helped this project, either through interview or the provision of family mementoes. The list is as exhaustive as I can recall, but many of the memories that are related come from years of casual discussions and friendship. The specific origins of these are neither remembered nor recorded.

Those friends who definitely contributed include:

Peter Small of Peat Inn whose encyclopaedic knowledge of early farm machinery was a tremendous help. He provided a great deal of material relating to sugar beet; he also kindly allowed me to quote from articles he had written on a range of subjects.

The Arbuckle family, from Lower Luthrie, for their father's papers, especially those relating to the closure of the sugar-beet factory.

Jimmy Garland, Cupar, who picked his way through a lifetime in livestock auction marts remembering the men and the issues of a generation ago.

David Leggat, Perth, of a younger generation, who added to my knowledge of the livestock auctioneering business and the characters associated with it.

Andy McLaren, Strathkinness, one of the oldest men I interviewed, but also mentally one of the sharpest. His father's diaries from the beginning of the century tell of country life long gone.

James McLaren, Cults, for papers belonging to his grandfather and his work in the war years.

Bill Brooksbank, Ladybank, for his voluminous memories of the potato industry and the changes in it.

Ian Waugh, Cupar, who told me of life and characters in the grain trade apart from himself.

Sam Anderson, Auchtermuchty, one of the last of the millers, who also related tales of the grain trade.

Andrew Peddie, Anstruther, who supplied me with copious notes of his own farming life and who then added anecdotes.

Maurice Milne, Pitscottie, for his recall of days in the Home Guard.

John Whitehead, Largo, who remembered the early days of the pea-vining group.

Jim and Ann Brand, Auchtermuchty, for their enthusiasm for the project and for their help in the early days.

Jack Roger, Boarhills. The first of the older generation I interviewed and one of the most helpful. Thanks also to his daughter, Joan.

Henry Watson, Anstruther, for allowing me access to his meticulous family records, for his photographs, and for being enthusiastic about the project.

Alistair Ewan, East of Scotland Growers, for his knowledge of vegetable growing.

Robin and Liz Lang, Pitscottie, for tales from their youth, some of which were too risqué to include in this book!

Andrew Logan, Cupar, who related tales of a farming life lived with enthusiasm.

Sandy Lathangie, Cupar, who recalled days of keeping pigs and also of breaking up hill ground in the post-war period.

Tom Pearson, Newburgh, who supplied me with many previously unpublished photographs.

Alan Clark, Windygates, for various articles about the early days of farming.

John and Doris Roger, Cupar, for their memories and for family papers belonging to George Watt.

John Purvis CBE, St Andrews, for taking time off from being a Member of the European Parliament to recall his youthful wartime days.

John Paul, Cameron, who also experienced these war years, and who remembered life as a farm worker in the pre-tractor era.

Bill Watson, Flisk, for his wide knowledge of the sheep sector.

Angus Hood, Alyth, for telling me about the early days on a berry farm and about the flax crop.

Will Robertson, St Andrews, one of the last of the horsemen, but still very active in life.

Jock McKenzie, Newburgh, one of the leading blacksmiths of his generation, for sharing some of his thoughts on working with horses.

Willie Porter, Carnoustie, for delving back into his roots and helping me find humour in the detail; also for organising a meeting with his uncle, John Gray, Carnoustie, another of the older generation, who helped by recounting his upbringing on a Fife farm.

John Cameron, Kilconquhar, whose enthusiasm for life should be bottled and then sold.

Robert and Betty Mitchell, Strathmiglo, who having spent much of their life milking cows, were able to tell me how it was in the milking byre.

Also John and Nancy Weir, Strathmiglo, who confirmed and added to these memories from the dairy sector.

Andrew Wilson, Newburgh, for his knowledge on horses and horsemen.

Pat Melville, Balmullo, for his family papers, including his mother's role in organising the Women's Land Army.

Pat Laird, Cairnie, for his memories of the seed trade.

Bob Steven, St Andrews for allowing me to view his family farm records.

DC Thomson & Co, Dundee for permission to reproduce photographs.

John MacNiven, St Andrews, for photographs from his family album.

Particular thanks to Kate Douglas, Gordon Berry and Dorothy Guyan for helping to sort out grammatical errors and to Posie Ridley for her enthusiasm: any remaining mistakes are my responsibility and not theirs. They also raised their eyebrows and a number of question marks when I delved too deeply into agricultural issues.

And finally, thanks to my brothers, John and Willie and sister Gina, who reminded me of some of the escapades and scrapes of an early life.

About the Author

Andrew Arbuckle comes from a Fife farming family which, like many others, could trace their roots back to the west of Scotland. The third son of a tenant farmer, Andrew went on to farm with his father in the 1970s. The farm was mainly arable, growing malting barley, seed potatoes and, before the closure of the local refining factory, sugar beet. The tenancy of the farm was given up in the late 1980s.

During his farming, Andrew contributed to a number of farming magazines and this interest developed into full-time agricultural journalism with the Dundee *Courier* where he was farming editor for fifteen years.

A short spell as a member of the Scottish Parliament followed between 2005 and 2007. Over the past twenty years, he has been a councillor in Fife. Nowadays, he is a freelance journalist combining this with agricultural public relations work.

He has two grown up daughters, one a solicitor and the other a psychiatrist.

Other Books from Old Pond Publishing

Farmer's Boy
Michael Hawker's detailed recollections of work on North Devon farms in the 1940s and 1950s. Paperback.

Farming Day by Day – the 1960s   JOHN WINTER
This selection from John Winter's reports in the Daily Mail vividly recalls the ups and downs of a decade when Britain's farmers were still being encouraged to produce more food. Hardback

Country Dance   HENRY BREWIS
Based in Northumberland, this grimly humorous fable from the 1990s tells the story of a hill farmer selling up and his farm being 'developed'. Paperback

Clarts and Calamities   HENRY BREWIS
The fictional diary of a year in the life of a Northumbrian hill farmer: 'the daily scribblings of an umpteenth-generation peasant'. Paperback

The Rural World of Eric Guy   JONATHAN BROWN
From the 1930s to the 1960s Eric Guy photographed the downland farming scene around his Berkshire base. Jonathan Brown has selected 174 of his most striking photographs and provides a knowledgeable text. Paperback

Farming & Forestry on the Western Front   MURRAY MACLEAN
This fine collection of fully captioned photographs from the First World War shows farming activities by the military including early tractors, horse-drawn binders and steam threshing. The second part of the book covers the work of Canadian and other allied foresters in France. Hardback

Free complete catalogue:

Old Pond Publishing Ltd, Dencora Business Centre,
36 White House Road, Ipswich IP1 5LT,
United Kingdom

Secure online ordering: **www.oldpond.com**
Phone: 01473 238200 Fax: 01473 238201